Designing for Older Adults

Human Factors & Aging Series

Series Editor:

Wendy A. Rogers, University of Illinois Urbana-Champaign

Given the worldwide aging of the population, there is a tremendous increase in system, environment, and product designs targeted to the older population. The purpose of this series is to provide focused volumes on different topics of human factors/ergonomics as they affect design for older adults. The books are translational in nature, meaning that they are accessible to a broad audience of readers. The target audience includes human factors/ergonomics specialists, gerontologists, psychologists, health-related practitioners, as well as industrial designers. The unifying theme of the books is the relevance and contributions of the field of human factors to design for an aging population.

Designing Telehealth for an Aging Population
A Human Factors Perspective
Neil Charness, George Demiris, Elizabeth Krupinski

Designing Training and Instructional Programs for Older Adults
Sara J. Czaja, Joseph Sharit

Designing Technology Training for Older Adults in Continuing Care Retirement Communities
Shelia R. Cotten, Elizabeth A. Yost, Ronald W. Berkowsky, Vicki Winstead, William A. Anderson

Designing for Older Adults
Principles and Creative Human Factors Approaches, Third Edition
Sara J. Czaja, Walter R. Boot, Neil Charness, Wendy A. Rogers

Designing Transportation Systems for Older Adults
Carryl L. Baldwin, Bridget A. Lewis, Pamela M. Greenwood

Designing Displays for Older Adults, Second Edition
Anne McLaughlin, Richard Pak

Designing for Older Adults: Case Studies, Methods, and Tools
Walter R. Boot, Neil Charness, Sara J. Czaja, Wendy A. Rogers

For more information about this series, please visit: www.crcpress.com/Human-Factors-and-Aging-Series/book-series/CRCHUMFACAGI

Designing for Older Adults

Case Studies, Methods, and Tools

Walter R. Boot
Neil Charness
Sara J. Czaja
Wendy A. Rogers

CRC Press
Taylor & Francis Group
Boca Raton London New York

CRC Press is an imprint of the
Taylor & Francis Group, an **informa** business

First edition published 2020
by CRC Press
6000 Broken Sound Parkway NW, Suite 300, Boca Raton, FL 33487-2742

and by CRC Press
2 Park Square, Milton Park, Abingdon, Oxon, OX14 4RN

© 2021 Taylor & Francis Group, LLC

CRC Press is an imprint of Taylor & Francis Group, LLC

Library of Congress Cataloging-in-Publication Data

Names: Boot, Walter Richard, author. | Charness, Neil, author. | Czaja, Sara J., author. | Rogers, Wendy A., author.
Title: Designing for older adults. Case studies, methods, and tools / authored by Walter Boot, Neil Charness, Sara J. Czaja, Wendy A. Rogers.
Description: First edition. | Boca Raton : CRC Press, 2020. | Series: Human factors and aging series | Includes bibliographical references and indexes.
Identifiers: LCCN 2020020825 (print) | LCCN 2020020826 (ebook) | ISBN 9780367220303 (hbk) | ISBN 9781138052857 (pbk) | ISBN 9781315167459 (ebk)
Subjects: LCSH: Human engineering. | Product design. | Older people--Services for.
Classification: LCC TA166 .B66 2020 (print) | LCC TA166 (ebook) | DDC 620.8/20846--dc23
LC record available at https://lccn.loc.gov/2020020825
LC ebook record available at https://lccn.loc.gov/2020020826

ISBN: 9780367220303 (hbk)
ISBN: 9781138052857 (pbk)
ISBN: 9781315167459 (ebk)

Typeset in Palatino
by Deanta Global Publishing Services, Chennai, India

Dedication

We dedicate this book to:

The older adults who have inspired us – our parents and grandparents

The participants in our research studies whose invaluable efforts have helped us to develop these guidelines

The countless students over the years who have made CREATE research possible

Contents

About the Authors

Walter R. Boot, PhD, is Professor of Psychology at Florida State University and Director of the university's Attention and Training Lab. He received his Ph.D. from the University of Illinois at Urbana-Champaign in Visual Cognition and Human Performance in 2007. Walter is one of six principal investigators of the multi-disciplinary Center for Research and Education on Aging and Technology Enhancement (CREATE), a long-standing and award-winning National Institute on Aging-funded center dedicated to ensuring that the benefits of technology can be realized by older adults. He is also Co-Director of the ENHANCE (Enhancing Neurocognitive Health, Abilities, Networks, & Community Engagement) Center, funded by the National Institute on Disability, Independent Living, and Rehabilitation Research, with a focus on how technology can support older adults living with cognitive impairment. His research interests include how humans perform and learn to master complex tasks (especially tasks with safety-critical consequences), how age influences perceptual and cognitive abilities vital to the performance of these tasks, and how technological interventions can improve the well-being and cognitive functioning of older adults. He has published extensively on the topic of technology-based interventions involving digital games. Walter is a Fellow of the American Psychological Association (APA) and the Gerontological Society of America and received the Springer Early Career Achievement Award from Division 20 (Adult Development and Aging) of APA in 2014 and the Earl A. Alluisi Early Career Achievement Award from Division 21 (Applied Experimental and Engineering Psychology) of APA in 2017.

Neil Charness, PhD, is William G. Chase Professor of Psychology, Director of the Institute for Successful Longevity, and Associate Director of the University Transportation Center (Accessibility and Safety for an Aging Population, ASAP) at Florida State University. He received his BA from McGill University (1969) and then his MSc and PhD from Carnegie Mellon University (1971, 1974), all in Psychology. Prior to coming to Florida State University, he was on the faculty at Wilfrid Laurier University and the University of Waterloo in Canada. Neil's current research focuses on human factors approaches to age and technology use, interventions to promote improved cognition, and aging driver and pedestrian safety. He is a Fellow of the APA, the Association for Psychological Science, and the Gerontological Society of America. Neil received the Jack A. Kraft Innovator award (with CREATE colleagues) from the Human Factors & Ergonomics Society in 2013. In 2016 he was awarded the Franklin V. Taylor Award for Outstanding Contributions in the field of Applied Experimental and Engineering Psychology from Division 21 of APA, the M. Powell Lawton Award for Distinguished Contribution to Applied Gerontology from Division 20 of APA, and the APA Prize for Interdisciplinary Team Research with CREATE colleagues. In 2018 he was awarded the title of Grandmaster of the International Society for Gerontechnology and APA's Committee on Aging Award for the Advancement of Psychology and Aging.

Sara J. Czaja, PhD, is a Professor of Gerontology and the Director of the Center on Aging and Behavioral Research in the Division of Geriatrics and Palliative Medicine at Weill Cornell Medicine. She is also an Emeritus Professor of Psychiatry and Behavioral Sciences at the University of Miami Miller School of Medicine (UMMSM). Prior to joining the faculty at Weill Cornell, she was the Director of the Center on Aging at the UMMSM. Sara received her PhD in Industrial Engineering, specializing in Human Factors Engineering, at the University of Buffalo in 1980. She is the Director of CREATE. Her research interests include aging and cognition, aging and healthcare access and service delivery, family caregiving, aging and technology, training, and functional assessment. She has received continuous funding from the National Institutes of Health, Administration on Aging, and the National Science Foundation to support her research. She is a Fellow of the APA, the Human Factors and Ergonomics Society (HFES), and the Gerontological Society of America (GSA). She is also Past President of Division 20 (Adult Development and Aging) of APA. She is also a past member of the National Research Council/National Academy of Sciences Board on Human Systems Integration. She served as a member of the Institute of Medicine (IOM) Committee on the Public Health Dimensions of Cognitive Aging and as a member of the IOM Committee on Family Caregiving for Older Adults. Sara is also the recipient of the 2015 M. Powell Lawton Distinguished Contribution Award for Applied Gerontology of APA; the 2013 Social Impact Award for the Association of Computing Machinery (ACM); the Jack A. Kraft Award for Innovation from HFES and the APA Interdisciplinary Team, both with CREATE; and the Franklin V. Taylor Award from Division 21 of APA.

Wendy A. Rogers, PhD, is the Shahid and Ann Carlson Khan Professor of Applied Health Sciences at the University of Illinois Urbana-Champaign. Her primary appointment is in the Department of Kinesiology and Community Health. She also has an appointment in the Educational Psychology Department and is an affiliate faculty member of the Beckman Institute, the Illinois Informatics Institute, and the Center for Social and Behavioral Science. She received her BA from the University of Massachusetts – Dartmouth and her MS (1989) and PhD (1991) from the Georgia Institute of Technology. She is a Certified Human Factors Professional (BCPE Certificate #1539). Her research interests include design for aging, technology acceptance, human–automation interaction, aging-in-place, human–robot interaction, aging with disabilities, cognitive aging, and skill acquisition and training. She is Director of the Health Technology Education Program; Program Director of CHART (Collaborations in Health, Aging, Research, and Technology; chart.ahs. illinois.edu); and Director of the Human Factors and Aging Laboratory (www.hfaging.org). Her research is funded by the National Institutes of Health (National Institute on Aging; National Institute of Nursing Research) and the Department of Health and Human Services (National Institute on Disability, Independent Living, and Rehabilitation Research). She is a Fellow of the APA, the GSA, and the HFES. She has received awards for her mentoring (HFE Woman Mentor of the Year, Fitts Education Award, APA Division 20 Mentor Award), her research (APA Division 21 Taylor Award, and with CREATE the APA Interdisciplinary Team and HFES Kraft Innovator Award), and her outreach activities (HFES Hansen Outreach Award).

chapter one

Introduction

The focus of this book is on designing for older adults. At the heart of good design for older adults is an awareness of, and adherence to, design principles that recognize the needs, abilities, and preferences of diverse groups of older adults. Comprehensive design guidelines are presented in our recent book, *Designing for Older Adults: Principles and Creative Human Factors Approaches*, Third Edition, and the reader is encouraged to consult that book (and other books in the Human Factors and Aging series) for valuable information regarding designing useful and usable systems for older users. While principles and guidelines are important, learning good design *as a process* is facilitated by seeing principles and guidelines in action, and by understanding how to use the methods and tools available to evaluate design. This book – *Designing for Older Adults: Case Studies, Methods, and Tools* – is intended to be a standalone book. The goal of this book is to provide illustrative "case studies" based on real design challenges faced by researchers of the Center for Research and Education on Aging and Technology Enhancement (CREATE) over the past two decades. These case studies are instances in which CREATE investigators have used human factors tools and user-centered design principles to understand the needs of older adults, identify where existing designs fail older users, and examine the effectiveness of design changes to better accommodate the abilities and preferences of the large and growing aging population. Presented case studies span many different domains including health, transportation, living environments, and communication and social engagement. Each chapter describes a method or tool vital to good design for older adults as well as important issues to consider in their use, specifically with older adults. After this description, relevant instances are presented in which the method or tool was successfully deployed, followed by a discussion of what was learned in terms of design. Chapter 8 presents an extended case study in which multiple methods and tools are applied to a single design challenge: the development of a computer system specially designed for older adults at risk for social isolation. To provide context for these case studies, methods, and tools, this chapter has three primary aims: (1) to present a conceptual model for design, (2) to discuss the importance of designing for older adults, and (3) to discuss characteristics of the aging population in terms of perceptual, cognitive, and motor abilities; diversity; and technology use, adoption, and preferences.

1.1 Conceptual Framework

Good design can be conceptualized as the process of producing an optimal fit between the demands of a system and the capabilities of the user.
The CREATE model (Figure 1.1) recognizes that the capabilities of different users vary, and that, on average, younger adults vary predictably from
older adults in important ways that impact their performance of tasks and
interactions with technology, including differences in perceptual, cognitive, and psychomotor capabilities. Attitudes and experience can differ
between younger and older adults as well. Design should accommodate
individual differences and variability in capabilities. Our framework
also recognizes that user–system interactions do not occur in isolation
but take place in a broader socio-cultural and physical environment that
can impact performance. For example, an older adults' community might
offer technology training classes that can increase the efficiency of an
older adult's interaction with a technology system without changing the
design of the technology itself.

This framework serves as a general guide for all of what CREATE
does with respect to design. The design process, including the methods
and tools described in this book, is aimed at understanding the demands
systems place on older users, mismatches between demands and capabilities, how redesign might produce a better demand–capability fit, and
how the environment might facilitate user–system interactions, all across
a variety of activity domains of everyday living. This process includes

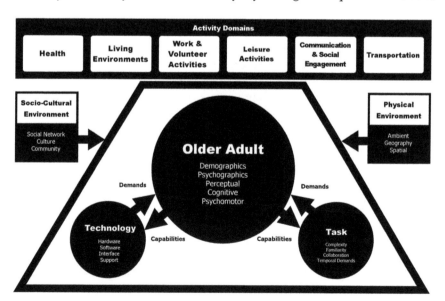

Figure 1.1 CREATE framework for a socio-technical system.

assessing older adults' needs and preferences, involving older adults in the design process, implementing usability methods, assessing design through simulation and performance modeling, and designing instructional support. Carefully following design guidelines and engaging in an iterative, user-centered design process helps ensure that the benefits of systems designed to improve productivity, well-being, health, and independence are accessible by all, regardless of age.

1.2   The Importance of Designing for Older Adults

The world is undergoing a major shift in demographics known as population aging. Changing demographics are primarily the result of two forces: declining birthrates and increasing longevity. In both industrialized and developing nations, both the number of older adults in the population and the proportion of the population that is made up of older adults are dramatically increasing and are projected to continue to increase. In the United States, between the years of 2017 and 2060, the number of people aged 65 and older is anticipated to nearly double, from 51 million to 95 million. The fastest-growing segment of the population is the "oldest old" (aged 85 years and older). Worldwide, the number of people 80 years of age or older is expected to triple to 426 million by the year 2050. The year 2018 marked a first in human history: for the first time, older adults (those aged 65 years and older) outnumbered children under 5 years of age. By the year 2035, in the United States, people 65 years of age and older will outnumber all people under the age of 18. That is, there will be a greater number of older adults than children for the first time in the nation's history. These facts highlight an important issue to consider in the design of systems: the population of users of systems now is different compared to what it was a few decades ago, and will be even more different in the coming decades.

Systems that do not consider the unique needs and capabilities of older users are likely to fail in their ability to support their use and adoption by a large and growing segment of the population. Designers need to consider that older adults will increasingly be buying and using their systems. In some cases, failure to consider the older user can result in less positive user experiences as well as lower adoption and use of systems or products by older adults. Further, unless older adults are involved in the design process, designers can incorrectly anticipate the needs and preferences of older users (out of tradition, the title of this book references the processes of "Designing for Older Adults," but "Designing *with* Older Adults" may be more appropriate). In some cases, lack of consideration for older users in the design process can result in system use that is slow and error prone. And for some systems, errors can have serious consequences, putting older users at greater risk for injury or death. This is especially true in the domains of healthcare and transportation. To ensure that systems are easy to use and useful for all users, designers must consider older adults as potential users of their systems. In addition to providing a better user experience for older adults, CREATE research has found that good design for older adults typically benefits younger users as well.

1.3   Understanding the Older User

Although there is substantial variability among older adults in terms of abilities, attitudes, experience, and preferences, and this variability should not be ignored, younger and older adults on average vary in predictable ways relevant to the design of systems. Although a complete review of differences is beyond the scope of this chapter, a few of the most relevant differences are highlighted below, along with examples of their implications for design.

1.3.1   Perceptual and Cognitive Abilities

As we age, our perceptual and cognitive abilities change, and these changes can have a large impact on how we interact with systems. Vision changes include changes in visual acuity (the ability to see details, particularly for near distances), decreased peripheral vision, slower visual processing, increased susceptibility to glare, and even changes in color perception. Age-related yellowing of the lens of the eye can make distinguishing between shades of blue, green, and violet more challenging. Hearing loss tends to increase with age, especially for higher frequency tones. Older adults can also experience greater difficulty understanding speech in noisy settings compared to younger adults. For example, older adults are more likely to have difficulty following a conversation in a noisy restaurant or understanding spoken dialogue in a movie in the presence of a background musical score. System design that does not account for age-related changes in vision and hearing, as well as other senses, can disadvantage older users. Examples include systems that feature small, low-contrast text and buttons and systems that provide auditory alerts using low volume, high-frequency tones.

Cognition (our ability to think, reason, and remember) also changes with age. In general, older adults tend to process information more slowly, have greater difficulty managing multiple tasks simultaneously, and sometimes have greater difficulty allocating cognitive resources efficiently in the performance of a task (attentional allocation). They also tend to take longer to learn novel skills, which can impact the amount of training and support necessary for the use of a system. However, the cost of new learning can be substantially less for tasks that are not novel, for example, learning a new version of a piece of software when one is already skilled at using the previous version.

Memory abilities change with age too, including working memory. "Working memory" refers to the ability to simultaneously store and manipulate information in mind, and this ability is important for problem solving, reasoning, and speech and language comprehension. Designs that place high working memory demands on the user, for example, by having them navigate complex menu structures to identify the most appropriate option, can negatively influence older adults' performance. Prospective memory, the ability to remember to do something in the future, can also be impacted. However, not all cognitive abilities show large age-related cognitive declines. Procedural memory (memory for the steps involved in how to perform a practiced task, the classic example being how to ride a bicycle) is often unaffected by age, and crystallized intelligence (knowledge about the world) remains stable or even increases later in life. Designs that take advantage of what older adults already know can help offset the impact of declines in other abilities. Memory limitations can be addressed by providing "knowledge in the world." This refers to having information (for example, about the sequence of actions involved in completing a task) displayed as part of the system itself, rather than relying on users to learn and remember these actions.

These are just a few of the important perceptual and cognitive changes related to advancing age that are relevant to system design. In this book, we focus largely on the design of systems for older adults experiencing normative age-related changes in cognition (changes most of us can expect to experience as we age). An even greater challenge is presented when considering how best to design systems for older adults experiencing cognitive impairment. This could be cognitive impairment as a result of mild cognitive impairment (MCI), stroke, traumatic brain injury because of a fall, or Alzheimer's disease. Although this book will not focus on the design of systems for older adults experiencing cognitive impairment, the same general principles and techniques can be applied.

1.3.2 Anthropometry, Movement Control, and Strength

In addition to perceptual and cognitive changes associated with aging, the body and physical capabilities can change with age as well. System design should consider age-related changes in physical dimensions (e.g., stature, weight), movement control, strength, and endurance. "Anthropometry" refers to the study of human-body dimensions, and anthropometric studies have found that older adults tend to be smaller in stature compared to younger adults, although there is variability with respect to how much, if any, change occurs longitudinally. These trends have important implications, for example, for the design of work and public spaces and the design of personal vehicles (e.g., placement of pedal controls).

Older adults are typically slower in their physical movements, and movements may be less precise and more variable for older adults compared to younger adults. The prevalence of tremor increases significantly with age. Among other implications, these changes have important implications for system "time out" parameters and the size of targets that older adults might need to hit with their finger when using a touchscreen display. These changes also have implications for the choice of a system input device and the parameters of that device (e.g., input gain). The generally slower walking speed of older adults compared to younger adults has implications for the design of pedestrian safety countermeasures. Muscular strength including grip strength also tends to decrease with age, and with age the likelihood of having arthritis also greatly increases. These changes can impact flexibility and the range of motion. Decreases in strength, flexibility, and range of motion can shape how users interact with certain systems and can differentially impact older users.

1.3.3   Technology Experience and Attitudes

The systems that designers are creating are often technological in nature, and designers need to consider that older adults often differ from younger adults in their experiences with and attitudes toward technology. These factors have profound implications for the design of systems and supporting materials (e.g., help manuals). There exists a pronounced age-related "digital divide" that older adults are less likely to use and own technology, especially newer technology, compared to younger adults. As of 2019 in the United States, 27% of older adults (those aged 65 years and over) report that they do not use the internet. In contrast, among the youngest adult cohort there is essentially universal internet use. For smartphone ownership, 47% of older adults do not own a smartphone, compared to only 4% of 18-to-29-year-olds (Figure 1.2). Similar age-related divides in technology use and adoption can be seen for social media, smart home devices, wearables, and gaming technology. Differences extend beyond use and ownership. CREATE research has found that even among younger and older adults who use and own technology devices, younger adults often demonstrate greater technology proficiency. Preexisting differences in technology experience and proficiency are important to consider in the design of technological solutions.

Why are older adults less likely to adopt and use new technology? To answer this question, we should first consider what motivates anyone to want to adopt new technology. Two of the most popular models to explain technology adoption are the Technology Acceptance Model (TAM) and the Unified Theory of Acceptance and Use of Technology

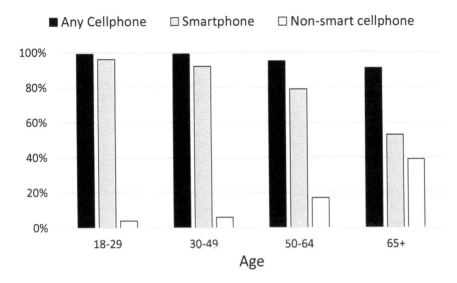

Figure 1.2 Cell phone ownership in the United States in 2019 as a function of age and cell phone time. *Source:* https://pewresearch.org/internet/fact-sheet/mobile/

(UTAUT) model. At the center of each model is the idea that technology adoption is primarily driven by perceived ease of use and perceived usefulness of the technology. That is, a piece of technology that is thought to help accomplish a goal and is anticipated to be easy to use is more likely to be accepted and adopted. Among non-internet-using older adults, some report that they believe the internet is too difficult to learn, and some report that they see no value in using the internet. This model fits well with why older adults choose to adopt technology, but the Senior Technology Acceptance Model (STAM; Chen & Chan, 2014) also recognizes age-specific factors. An older adults' cognitive and physical health, as well as their technology self-efficacy and anxiety, all play important roles in STAM in determining whether a specific piece of technology is adopted. Within CREATE's own model (Czaja et al., 2006), we also find that these factors are important. In terms of technology attitudes, older adults still report less comfort using technology as well as less technology self-efficacy compared to younger adults, although in general over the past 20 years older adults have reported more positive technology attitudes (Lee et al., 2018).

Any designer of technology solutions for older adults needs to consider that older adults' technology experience, proficiency, and attitudes are different than younger adults', and successful design of devices and supporting materials need to consider these differences. Not considering age-related changes can suppress technology adoption and result in more negative user experiences.

1.3.4   The Importance of Considering Variability

Although older and younger adults vary in ways that are predictable, it is important to understand that older adults are not a homogenous group. In our research, we have tested older adult participants with faster processing speed scores and higher working memory capacity than some younger adults. We have also observed lower mobile device proficiency for younger adults compared to some older adults. Even among older adults, technology proficiency tends to decrease with age, and technology ownership and access vary as a function of socioeconomic status and location. There are two important implications of this variability among older adults with respect to design. The first, as already alluded to, is the fact that good design for older adults also often benefits younger adults. The second is that, when evaluating system usability and the effect of various design changes, it is important to capture this diversity in study samples if the system is intended to be used by a broad range of diverse older users.

1.4   Preview of Case Study Examples

The goal of this book is to provide insight into the process of designing for older adults through case study examples. How would one design a computer system for older adults who have minimal previous technology experience? How would one effectively test a roadway countermeasure to protect older adults from rare but serious automobile crashes? How do we design systems with the appropriate types and amount of support? And how do we understand the needs of older adults in their daily lives to begin with? Chapters will focus on methods to define older adult groups based on the research questions under investigation: how to assess the needs of older adults, how to evaluate usability, and how to design instructional support for systems. Additionally, chapters will explore the potential of simulations and performance modeling to inform design decisions for older adults. Subsequent chapters will first introduce a research tool or method, followed by multiple real examples of how CREATE has used those tools and methods to solve a design challenge. Multiple case studies per chapter will provide these insights across a variety of domains including health, transportation, living environments, and communication and social engagement.

1.5   Conclusions

Due to demographic changes, older adults will make up a larger proportion of the population in the future, and age is associated with a variety of perceptual, cognitive, and physical changes that have implications for system design. Age-related technology adoption and proficiency differences are evident as well. If these differences are not accounted for, older users can be disadvantaged. Their use of a system might be slower and more error prone compared to younger adults, and poor design of safety-critical systems can put older users at greater risk for injury or death. The aim of this book is to provide guidance on how to design for older adults by providing case study examples of design methods and tools. While design principles and guidelines are important, learning good design as a process is facilitated by seeing principles and guidelines in action and understanding how to use the methods and tools available to evaluate design.

References

Chen, K., & Chan, A. H. S. (2014). Gerontechnology acceptance by elderly Hong Kong Chinese. A senior technology acceptance model (STAM). *Ergonomics*, 57(5), 635–652. doi:10.1080/00140139.2014.895855.

Czaja, S. J., Charness, N., Fisk, A. D., Hertzog, C., Nair, S. N., Rogers, W. A., & Sharit, J. (2006). Factors predicting the use of technology: Findings from the Center for Research and Education on Aging and Technology Enhancement (CREATE). *Psychology and Aging*, 21(2), 333–352. doi:10.1037/0882-7974.21.2.333.

Lee, C. C., Czaja, S. J., Moxley, J. H., Sharit, J., Boot, W. R., Charness, N., & Rogers, W. A. (2018). Attitudes toward computers across adulthood from 1994 to 2013. *The Gerontologist*, 59(1), 22–33. doi:10.1093/geront/gny081.

Additional Recommended Readings

Baldwin, C. L., Lewis, B. A., & Greenwood, P. M. (2019). *Designing transportation systems for older adults*. CRC Press.

Cotten, S. R., Yost, E. A., Berkowsky, R. W., Winstead, V., & Anderson, W. A. (2016). *Designing technology training for older adults in continuing care retirement communities*. CRC Press.

Czaja, S. J., Boot, W. R., Charness, N., & Rogers, W. A. (2019). *Designing for older adults: Principles and creative human factors approaches* (3rd ed.). CRC Press.

Czaja, S. J., & Sharit, J. (2016). *Designing training and instructional programs for older adults*. CRC Press.

Demiris, G., Krupinski, E., & Charness, N. (2011). *Designing telehealth for an aging population: A human factors perspective*. CRC Press.

McLaughlin, A., & Pak, R. (2020). *Designing displays for older adults* (2nd ed.). CRC Press.

chapter two

Defining Older Adult User Groups

By definition (Merriam-Webster, 2003) a group is "a number of individuals who have some unifying relationship." In the design world, we consider a user group to be a group of people who use or will use some product, such as a smartphone, software application (e.g., Microsoft Word), or a service, such as an exercise program provided at a local senior center. User involvement in design is an essential component of the user-centered design approach. Lack of attention to user needs and preferences can lead to a lack of adoption or abandonment of a product, errors, inefficient use of a product, and user dissatisfaction. For example, the IBM PC Jr. was designed in the 1980s to be used in the home. However, the product turned out to be a commercial failure and was actually pulled from the market due to a number of design and implementation problems. One problem was the small "chiclet" keyboard, which was difficult to use for extended keying, such as word processing. This design flaw caused frustration and dissatisfaction as users expected that they would be able to use the PC Jr. at home for this type of activity (Sanger, 1984). There were also problems with the amount of memory storage, which was insufficient for some of the desired applications. Generally, design problems such as these can be avoided if user needs, abilities, and preferences are accounted for during the design process.

Including users in the design process involves identifying user groups; understanding their needs, characteristics, and preferences; and involving representative users in the actual design and evaluation of products and systems. In this chapter, we discuss older adult user groups and provide a summary of the vast sources of diversity among older adults. Our goals are to discuss the importance of considering older adult user groups and to describe some important considerations when selecting and involving older adults in the design process.

2.1   Older Adult User Groups

As noted elsewhere in this book, the number of older adults in the U.S. population and worldwide is growing at an unprecedented rate (see Chapter 1). Older adults are an important user group and are active consumers of products and users of systems. For example, the use of the internet and mobile devices is increasing among adults aged 65+, and many older adults are actively engaged in employment, learning, wellness, and leisure activities. As shown in Figure 1.1 (Chapter 1), engagement in these activities typically involves interactions with products, devices, and technologies. Unfortunately, all too often, aging adults are not considered as a valid user group in the design of products and technology applications. The result is products and technologies that are difficult for many older people to use.

In the pilot testing of our PRISM trial that evaluated a software system designed for seniors (discussed in Chapter 8), we found that the majority of our participants had difficulty using the computer mouse. Participants required extra training and the elimination of the need to double click to be able to use the system. Had we not pilot-tested our training protocol with older adults, we would not have become aware of this issue, which in turn would have resulted in problems using the PRISM software. Furthermore, these problems may have been erroneously attributed to the design of PRISM rather than to older adults' challenges of using the mouse.

2.2   Diversity of the Older Adult Population

Including older adult user groups in product design and testing necessitates understanding the characteristics of this population. A common myth about older adults is that they are all alike – with one defining characteristic, namely "old." In fact, older adults are diverse in many dimensions, which is another important consideration for design. Aging is a continual process that is influenced by a variety of factors such as genetics, lifestyle, and environment; thus people age differently.

There are of course normative age-related changes such as declines in sensory processes, strength, agility, and some aspects of cognition. For example, older adults generally experience declines in vision, audition, memory, attention, and processing speed (see Czaja, Boot, Charness, & Rogers, 2019, for a more complete discussion of these issues). These age-related changes do not reflect changes due to a disability or a disease but are a normal part of the aging process and have clear implications for the design of products, systems, training programs, and instructional materials. Older adults with a chronic illness such as dementia or severe arthritis or those who are aging with a disability have unique characteristics and needs. If these subpopulations of older adults are potential users of a product or system, they must also be included in user testing. For example, we are currently evaluating whether the PRISM system is usable for people with mild cognitive impairment (MCI). People with MCI were not included in our initial PRISM usability studies, and thus we cannot assume that PRISM will be usable by this subpopulation of older adults. Future generations of older adults will be better educated, more active, and healthier on a number of dimensions than prior generations of older people. Aging is also associated with plasticity such that older people can learn new things and experience gains in functional capacity.

With respect to variability among older adults, there are within-person differences, which are generally referred to as *intra-individual differences*. These are differences in a person that occur over time. For example, the performance of an individual might be measured every five or ten years, and there would likely be changes in variables such as knowledge, skill level, or physical dimensions. There can also be variability within an individual's performance across days or even within a given measurement occasion due to factors such as fatigue, acute illness, distractions, medication, or attention lapses. Think, for example, of someone who is under stress or who has had little sleep. Their performance on a task may not be at their optimum level.

Inter-individual differences refer to differences between people. These differences can occur between age groups – for example, comparisons between younger and older people – as well as within an age group. Older adults vary in a variety of dimensions. One important source of variability among older adults to consider is *age*. There are differences between adults in their 60s and 70s and those who are in their 80s and 90s. People in the latter cohorts are more likely to be frail, to have chronic conditions, and to be socially isolated than older adults in their 60s. For many products and systems, consideration of cohort differences in needs and abilities is important as people who are in their 80s and 90s will interact with a variety of products and devices including technology applications such as monitoring and wearable devices, and will represent a significant portion of the population in the upcoming decades.

Gender should also be considered in the design process. There are clear physiological and anthropometrical differences between men and women. Older women tend to be smaller in stature and have less strength than older men. Chronic conditions such as osteoarthritis are more common among older women than men. There are also gender differences in needs, preferences, life experiences, and attitudes.

Culture/ethnicity is another source of individual differences. The older population in the United States is becoming more ethnically and culturally diverse. In the upcoming years, we will see a growth in the proportion of Hispanic, African American, and Asian older adults in the United States (see Figure 2.1). Individuals from ethnic backgrounds other than non-Hispanic whites have different life experiences, attitudes, and beliefs. They are also less likely to have English as their primary language. Today, system designers are beginning to recognize these differences, and many

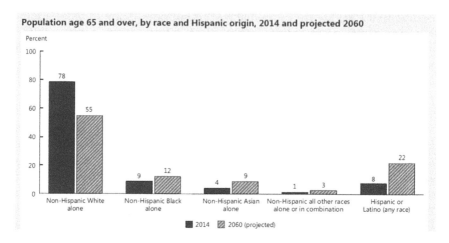

Figure 2.1 Increasing ethnic/cultural diversity of the older adult population. *Source*: Federal Interagency Forum on Aging Related Statistics (2016)

products and systems such as ATMs and ticket kiosks now include multiple languages.

Older adults vary widely in *socio-economic status (SES)*, which typically reflects race, income, and educational attainment. The ability to purchase new products is impacted by SES. Thus, older adults from lower SES brackets may have less access to and limited experience with newer technologies. Thus, SES is an important variable to consider when involving older adults in user testing sessions.

There are vast differences in *living environments and contexts* of aging adults. The majority of older adults live in the community in a variety of dwelling types including single houses, apartments, senior living facilities, and condominiums. A small percentage of older people live in assisted living or skilled nursing facilities. Clearly, the type of living environment has an impact on access to community events, educational programs, and utilities such as broadband. Currently about 25% of older adults live in small towns or rural locations, where access to the internet is often limited. Access to transportation and other services can also be challenging, which may make it difficult for these individuals to access needed services and to participate in programs such as those offered at senior centers.

Skills and *skill attainment* vary among older adults due to a variety of factors such as interests, SES, or employment factors. We evaluated a training program aimed at teaching basic computer skills to older adults. The entry skill level of the participants varied, which in turn had an impact on people's experiences with the program. Those with lower entry-level skills expressed a need for more practice and a slower classroom pace, whereas those with higher entry-level skills felt that the pace of the class was too slow and were bothered by repeated questions from the less skilled attendees. Generally, it is best not to mix people of varying skill levels in a technology-focused training or instructional program (see Chapter 7).

Older adults also vary in terms of attitudes and preferences, as do individuals of other age groups. It is commonly assumed that older people are "technophobic" and therefore unwilling to engage in new learning around technology. We have found the opposite in our research. In fact, older people are interested in and willing to interact with technology if they perceived it as valuable, useful, and usable, and if the appropriate instructional and technical support is provided. In fact, a recent analysis of attitudinal data collected by CREATE over the last two decades (Lee et al., 2019) has shown that although age differences in attitudes toward computers remain, they are becoming more positive among recent cohorts of older adults (see Figure 2.2).

Other important sources of variability among older people include *general literacy, health literacy,* and *numeracy. General literacy* is defined as the ability to use and comprehend print and written information. Within this category, there is a distinction between prose literacy and document literacy. *Prose literacy* refers to the knowledge and skills required to use and understand information from text, newspapers, package instructions, or the content of web pages. *Document literacy* refers to the knowledge and skills needed to use information contained in documents such as job applications, benefit forms, transportation schedules, or maps. Unfortunately, a large number of older adults have low literacy and therefore have difficulty understanding text and written instructions.

Health literacy is a distinct type of literacy and refers to the ability to obtain, process, and understand health information. This concept includes medication information, medical websites, brochures, consent forms, and communications from healthcare professionals. Unfortunately, like general literacy, health literacy is low among aging adults. This is a serious concern given the increased need for older adults to interact with systems such as electronic health records (EHRs). Instructions to operate medical equipment often require high levels of both general literacy and health literacy.

Numeracy refers to the ability to use numbers, understand mathematical concepts, and apply these to solve a variety of problems. Numeracy impacts important life activities such as financial and health management. For example, the ability to understand graphs and charts depends on numeracy. Similar to health literacy, numeracy tends to be at low levels in older adults, especially among those in lower SES strata. In the case study presented below, we discuss how low levels of numeracy affected the ability of middle-aged and older adults to use the patient portal of an EHR.

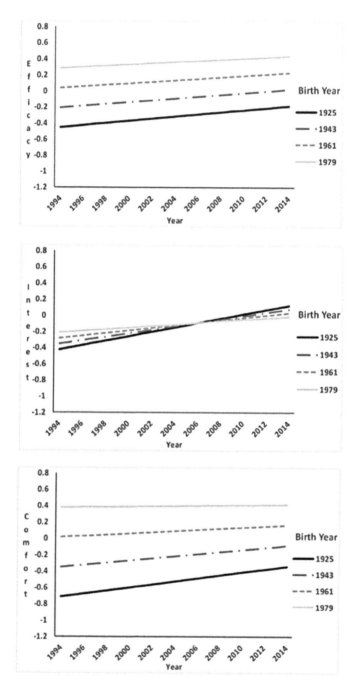

Figure 2.2 Cohort differences in comfort with computers. *Source*: Lee et al. (2019)

Health and functional status vary significantly among aging adults. Although, as noted earlier, current generations of older adults are healthier than those of previous generations on some markers of health, many older people have one or more chronic conditions such as arthritis, hypertension, or diabetes. They are more likely to have problems with mobility. Ethnic minorities are more likely to report worse health than non-minorities and to have higher rates of chronic conditions such as hypertension, diabetes, and Alzheimer's disease. Women are more likely to have arthritis than men.

Functional status includes the ability to perform *activities of daily living* (ADLs) such as bathing and eating and *instrumental activities of daily living* (IADLs) such as managing money and medications. These activities are essential to independent living, and many older adults have difficulty performing one or more of these activities. The prevalence of difficulties performing ADLs and IADLs increases with age and with the development of chronic conditions such as arthritis or cognitive impairments. *Enhanced activities of daily living* (EADLs) refer to activities that involve new challenges or new learning, such as learning to use a new mobile device, online banking, shopping software, or medical devices. EADLs are critically important in today's technologically driven world. People are constantly confronted with the need to learn new technologies and new ways of performing activities.

Older adults vary in terms of their *sexual orientation* and *gender identity*. The number of lesbian, gay, bisexual, and transgender (LGBT) older adults is increasing and will continue to grow in the upcoming decades. These populations of older adults confront special challenges, such as differential access to healthcare, housing, and legal protections. LGBT older adults are more likely to deal with problems of depression, diabetes, and hypertension than are their heterosexual and cis-gender counterparts, and often have less access to needed caregiver programs and supports. Problems with social isolation are common among LGBT seniors.

2.3   Considering Diversity of the Older Adult Population in Design

In the preceding section, we provided an overview of the sources of dimensions on which older adults vary. Our intent was to highlight the diversity of the older adult population and the importance of considering diversity in design. As discussed, older adults vary on a number of dimensions including physical, socio-demographic, attitudinal, prior experience, and preferences. Accommodating older adult user groups requires understanding these sources of variability in user characteristics and a user-centered design approach that includes representative samples of older adult user groups. For example, when airbags were first deployed in automobiles they had only been tested on large male dummies, putting smaller passengers such as females and children at risk of injury. Over time, with reports about the safety concerns associated with airbag design, dummies of female passengers were also included in the testing of airbags. As illustrated by this and many other examples, the involvement of users in design generally leads to more effective, more efficient, and safer products and systems, as well as to higher user satisfaction and uptake of products.

For many systems and products, user groups are broad and extend beyond older adults. If we think, for example, about a home monitoring system, the user groups are likely to include the older adult, family members, informal care providers, and formal care providers. When considering the design of medical products, users are likely to include healthcare professionals of varying skill levels, informal caregivers (e.g., family members), and patients. Our emphasis in this chapter is on older adults, as we want to ensure their inclusion in the design process. Identifying potential user groups is an important part of the design process and many times requires "thinking outside of the box." It is generally a good idea to assemble people with differing backgrounds relevant to the product or system to brainstorm about the characteristics and needs of various potential user groups. In the design of a medical device, for example, the list of users to consider might include both formal and informal care providers and patients. We evaluated a telemedicine system for frail older adults with hypertension (Czaja, Lee, Arana, Nair, & Sharit, 2014). The system was installed in the participants' homes and sent health-related data such as blood pressure and weight measurements to nurses at a service agency. In the design of the system we had to consider the usability requirements for both the older adults (e.g., ability to read the display, understand how to accurately measure blood pressure) and the nurses (e.g., scheduling and frequency of data transfers, highlighting important information).

2.4   Case Study 1: Electronic Health Records (EHRs)

2.4.1   Research Questions

Electronic health records are becoming increasingly integrated into the healthcare environment. EHRs are digital versions of a patient's paper health record and contain the medical and treatment history of patients including diagnoses, medications, treatment plans, laboratory and test results, and immunization dates. The information can be shared among authorized providers including all of the clinicians involved in a patient's care. EHRs are intended to automate and streamline provider workflow, improve care coordination, and ultimately improve patient care (Office of the National Center for Health Information Technology, 2019).

Patient portals, often referred to as *patient health records* (PHRs), are secure online websites that allow patients to access personal health information from anywhere with an internet connection. Patients can view health information such as recent provider visits, medications, and lab test results. Many portals also allow patients to communicate securely with their doctors, schedule non-urgent appointments, view educational materials, and refill prescriptions. Patient portals are intended to improve the efficiency of the healthcare process (e.g., ease of making appointments), facilitate communication between the patient and the provider, and enhance a patient's involvement in and management of their health.

Given the increasing proliferation of PHRs, we evaluated the degree to which older adults could use these systems and how user characteristics, such as computer literacy, health literacy, and numeracy, influenced user performance of health-related tasks. The focus on older consumers is particularly important given that their greater propensity toward illnesses and disabilities means they use more medical care services than other age groups. Further, as discussed earlier, older adults are heterogeneous and thus we need to understand how individual difference factors impact the use of a PHR.

2.4.2   Examining How Individual Characteristics Influence Use of a Patient Portal

We undertook a series of studies to examine the influence of individual characteristics on the use of a patient portal. In our initial study, we conducted focus groups to examine patients' perceptions of PHRs (Zarcadoolas et al., 2013). The participants included ethnically and culturally diverse adults from lower educational backgrounds, all of whom had some experience with computers. Our goal was to identify the perceived utility of the portals and barriers to their use. Participants were introduced to the concept of a portal and shown video demonstrations. A trained moderator then asked the participants a series of questions related to the perceived value and usability of the system. Most participants reported that they perceived value in using a portal for activities such as making appointments or accessing medical records and lab test results. However, the participants also reported that the reading and health literacy requirements seen in the video examples seemed challenging. Most participants were frustrated by the highly technical language and found it difficult to understand the graphs and charts. Many expressed concerns about their inability to interpret the information correctly. They indicated that they would use a portal if available and if they had adequate support.

In a second study (Czaja et al., 2015), we used task analysis (see Chapter 4) and literacy load analysis (examines literacy requirements such as required reading grade level and sentence complexity) to identify the demands associated with using a PHR. We also conducted a usability study to examine challenges encountered by adults from lower SES strata and with low health literacy using PHRs to perform health management activities. We evaluated three widely used PHRs and focused on five basic tasks: finding basic health information, medication management, interpreting lab/test results, health maintenance/preventive care, and communication/appointment setting.

The literacy load analysis indicated that all of the three systems had a high literacy load, meaning that the text was likely to be difficult to read and understand. For example, the use of complex medical technology was ubiquitous across all three systems. Participants had to interpret highly technical names of lab and screening tests. Labeling also presented barriers, as some of the labels were not commonly used terms, were not logically linked to associated content, or were inconsistent. The task analysis revealed that all of the portals placed demands on cognitive abilities such as memory and attention (see Figure 2.3). Using the PHRs to complete the tasks also required health literacy, numeracy, and basic computer, mouse, and window skills. The usability assessment indicated that participants experienced difficulty navigating through the systems to find information, interpreting lab and test results, and using the portals to complete tasks such as refilling prescriptions. The participants did, however, indicate that they perceived PHRs as valuable and that they would be helpful in terms of health management.

In a third larger study (Taha, Czaja, Sharit, & Morrow, 2013), we evaluated the ability of middle-aged adults and older adults to use a simulated PHR to perform 15 common health management tasks encompassing medication management, review/interpretation of lab/test results, and health maintenance activities. Due to the experimental nature of the study, we used a simulated PHR, as we could not provide participants with access to protected health information. The simulation was modeled after a widely used PHR system, Epic's *MyChart*, which we populated with fictitious patent data (see Figure 2.4). Our data showed that overall, the participants performed poorly on the tasks such as interpreting risk information from graphs, evaluating whether values were out of "normal" range, understanding dosage information, and finding the date and time of an upcoming appointment. The data also showed that education, internet experience, cognitive abilities, numeracy, and age were significant predictors of performance. In terms of cognitive abilities, verbal ability was the most significant predictor, followed by reasoning and executive functioning. Overall, participants who were older, less educated, had lower cognitive abilities, numeracy, and limited internet experience performed less well on the tasks.

Subtasks/steps	Sensory/Perceptual Demands	Cognitive Demands	Response Execution Demands
Access information on lab test results	Vision and perceptual recognition: • Locate/Recognize "Request your PHR" button on portal home page **or** • Locate/Recognize "Lab/Diagnostic Reports" from "Review" subheading on the sidebar	Selective attention, text and health literacy/comprehension, working memory, long term memory • Discriminate among information on page and ignore irrelevant information • Read and comprehend text on page • Remember and comprehend that box labeled "Request Your PHR" has lab information • Read and comprehend text on side bar • Remember/comprehend that "Lab/Diagnostic Reports" contains lab/test result information • Understand and remember basic mouse and window operations	Fine motor skills; manual dexterity: Use of mouse: • Cursor positioning • scrolling; • clicking
If Selected "Request your PHR"	Vision and perceptual recognition: • Locate/identify "results" box on screen	Selective attention, text and health literacy/comprehension, working memory, long term memory; • Discriminate among information on screen and ignore irrelevant information • Read and comprehend information on screen • Understand and remember basic mouse and window operations (e.g., scroll to get information on lab/test results)	Fine motor skills; manual dexterity: • use of mouse; scrolling; clicking
Access the correct lab/test results cholesterol value –	Vision and perceptual recognition: • Locate /identify cholesterol	Selective attention, text and health literacy/comprehension, working memory, basic numeracy • Read and comprehend labels in results box; • Discriminate among the types of test values presented; • Ignore irrelevant information on the page; • Identify numbers	
Determine if value within normal range	Vision and perceptual recognition: • Locate/identify the "result(normal range)" information	Selective attention, text and health literacy/comprehension, working memory, long term memory, basic numeracy, analytical numeracy • Read and comprehend labels • Discriminate among the type of value information presented (e.g., attribute vs. result) • Compare values (your value vs. your normal range) • Understand meaning of mathematical symbols - parentheses • Understand and make equivalences – does value fall within higher range	

Figure 2.3 Example of the task analysis for interpreting lab/test results for cholesterol from a personal health record. *Source*: Czaja et al. (2014)

2.4.3   Design Value

The data from our studies clearly illustrate the importance of examining individual differences when evaluating a product or system as it points to which groups of consumers are likely to have difficulty using that product or system. Our series of studies demonstrate the value of including users in the design process as they provide valuable insights into aspects of a design that may prove challenging as well as aspects that are beneficial. Users may also provide suggestions regarding how a design might be improved. A design team that is quite familiar with a product or system may have been unaware of factors until being presented with the additional insights provided by users. It is challenging for designers to identify all possible errors a user may commit or other ways of interactions with a system that could go wrong.

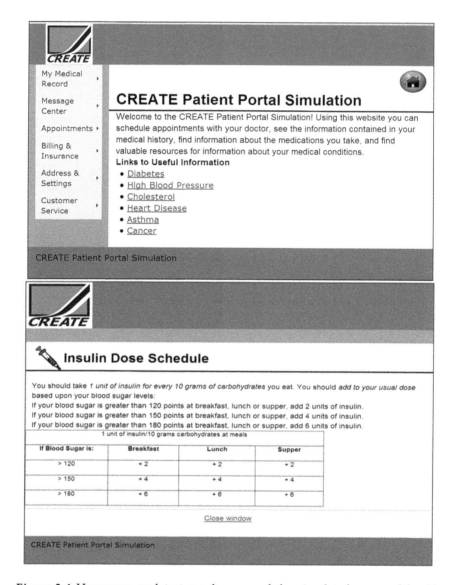

Figure 2.4 Homepage and test results page of the simulated personal health record. *Source*: Taha et al. (2013)

In the case of a PHR, our findings suggest that many consumers would have difficulty using a PHR. For example, current PHRs require high levels of general literacy, health literacy, and numeracy. To successfully use a PHR, users must understand complex technical and medical language as well as have the ability to understand mathematical information such as interpreting graphs and charts and identifying values that are out of range. These tasks can be challenging for many users, but, as noted, general literacy, health literacy, and numeracy are low among many older adults. From a design perspective, it would be useful to examine alternative ways of presenting graphical information or to provide some type of aid to facilitate interpretation. It might also be helpful to use highlighting techniques to indicate lab or test values that fall outside the normal range and to provide some type of explanatory information around those results.

Labels of menus and submenus need to be consistent and clearly related to the content, and systems need to be easy to navigate. In this study, internet experience was a significant predictor of task performance, and it was harder for people with limited or no experience to use the systems. This finding points to the need for training and the value of making system navigation less complex. We cannot assume that all system users will have internet experience, especially those in the older cohorts. Finally, the text of PHRs needs to be simpler and less technical. Individuals with low general literacy and low health literacy find it difficult to understand highly technical medical terms. Verbal ability and education were significant predictors of performance; people with lower verbal ability and education performed less well. This finding points to the need for using simpler language and providing more supporting information to help users understand what a term means. Although overall, most participants in our studies found the PHR to be valuable and stated that it would be helpful with respect to health management, they also indicated challenges and barriers to use. This information is valuable for the designers of these systems.

2.5   Case Study 2: A Technology-Based Caregiver Intervention Program

2.5.1   Research Questions

The prevalence of Alzheimer's disease (AD) is expected to increase significantly in the United States and worldwide in the coming decades, with a related increase in the burden on the patient, family members, and society. AD is a devastating disease that erodes the quality of life and functioning of the individual with AD, generates high levels of stress and burden on family caregivers, and results in substantial economic burdens to society. The majority of people with Alzheimer's disease are cared for at home by family members or "fictive" family members such as friends or neighbors. Although there are positive aspects of caregiving such as giving back to someone and experiencing personal growth, the negative consequences of caregiving are well documented, especially for caregivers of patients with AD. Caregiving can be extremely stressful and result in negative physical and emotional health consequences for the caregiver, disruptions in social and family relationships, and economic burdens.

A broad range of intervention studies have been aimed at the caregiver, and, overall, recent summary analyses have shown significant effects in reducing burdens, lowering depression, and delaying the placement of patients. However, for a variety of reasons such as logistic issues, many caregivers are unable to avail themselves of these programs. We evaluated whether we could deliver an evidence-based intervention to family caregivers of patients with AD using technology. The medium we chose for delivering this intervention was a screen phone, which at the time the research was conducted was the least cumbersome option available with respect to combining text and speech information, and represented a familiar technology (the telephone) to our caregivers who were largely from lower SES strata.

Our research questions centered on examining the usability and acceptability of the technology system and determining whether the intervention resulted in positive outcomes for the caregivers. We had to consider how to accommodate individual differences in our design as our user population included Hispanic, black/African American, and Haitian caregivers. Our user group included the interventionists, trained psychologists, and social workers who delivered the intervention. The study was a randomized pilot trial with three conditions: (a) the "VideoCare intervention condition"; (2) an "attention control condition"; and (c) a "usual care control condition." Participants in the attention control condition received a nutrition intervention (Czaja, Loewenstein, Schulz, Nair, & Perdomo, 2013).

2.5.2 Designing the VideoCare System to Accommodate Individual Differences

We chose the Cisco video telephone system as the mechanism of intervention delivery. The videophone technology allowed face-to-face communication between the interventionists and the caregivers, and allowed the caregivers to participate in face-to-face online support groups from the comfort of their own home. In addition, the caregivers had access to a broad array of intervention features such as a resource guide, educational videos, and caregiver tips. The features were presented in hierarchal menus in a multi-modal format (text and speech). The menus contained submenus (see Figure 2.5). For example, the resource guide contained information on respite care, legal services, meal services, etc. Caregivers navigated through the menus using the numeric keypad on the videophone. Upon login, the system automatically switched to the language preference chosen by the participant. The videophone was installed in the homes of 60 caregivers assigned to the intervention condition, and

Figure 2.5 Homepage of the VideoCare menu. *Source*: Czaja et al. (2013)

linked via a DSL connection to a secure server at the host site (University of Miami). Caregivers were individually trained to use the system and were provided with a help card and access to technical support.

The intervention was designed to (a) provide caregivers with access to information about dementia, caregiving, and community resources; (b) enhance knowledge and caregiving skills; (c) enhance social support; and (d) reduce barriers between caregivers and healthcare professionals. The intervention was delivered over a five-month period and was tailored to the specific needs and language preference of the caregiver.

One design challenge was accommodating the language preference of the caregivers. As noted, our sample included Hispanic and Haitian caregivers, many of whom were non-English speakers. To ensure our target population had access to the intervention, we had to translate all aspects of the intervention (all features and material within features), the help materials, and the assessment instruments into Spanish and Haitian Creole. This proved challenging as Creole is difficult to translate into written text. In addition, the intervention material was in a multi-modal format, namely, text screen information accompanied by speech. We thus had to hire a Creole-speaking translator as well as Creole-speaking interventionists. Translation of materials involves a forward- and backward-translation process, and thus was very time consuming.

A second challenge was accommodating the literacy levels of our participants, as many had very low general literacy. We had to ensure all of our text was at the sixth-grade reading level. In addition, some of our Haitian participants were unable to read, so we had to increase our use of speech as an interface to accommodate this population.

A further challenge was fostering acceptance of the technology as a viable tool for intervention delivery among our interventionists. They had previously delivered the intervention face-to-face and had some skepticism about fostering a bond with the caregivers via videoconferencing. We also needed to facilitate communication among our technical team, clinical team, and vendors to ensure that they were "all on the same page" with the intervention and were cognizant of each other's needs and preferences. For example, the interventionists were not aware of some of the constraints confronting the system programmers, and the programmers were unaware of the limited technical skills of some members of the technical team. Finally, we encountered some challenges with the living environments of our participants. Unfortunately, we made an erroneous assumption that the caregivers would have internet in their homes. In fact, none of the caregivers had access to the internet, and thus we needed to work with the local communications companies to make the internet available to them.

Prior to implementation we pilot tested the system and discovered aspects of the system that needed refining. These factors included issues such as the clarity and pace of the speech, the protocol for logging onto the support groups for both the caregivers and the interventionists, and the need for an easy-to-use graphical help card. The interventionists also needed to be trained in the installation protocol, which was initially challenging, as well as on how to troubleshoot basic system failures.

Overall, we found that the intervention was feasible, acceptable to both caregivers and the interventionists, and resulted in positive outcomes for the caregivers. We were able to train all of the participants on the use of the system, which was an accomplishment, as most of our caregivers had limited experience with technology. The caregivers indicated that they found the system easy to use and valuable. They particularly enjoyed participating in the online support groups as they could join in the groups without having to leave their homes. In addition, we found that caregivers who received the intervention reported less burden and more social support, and felt more positive about the caregiving experience.

2.5.3 Design Value

The VideoCare project clearly demonstrates the importance of considering varying user groups and understanding their needs, preferences, and abilities. In this case, our user groups included both caregivers and interventionists. The system had to be designed so that both groups of users were able to successfully use the system. As such, we included both groups of users in our pilot testing and learned that caregivers and interventionists experienced challenges with the initial design of the system. As a result, we modified the protocol for the focus groups to make the login protocol simpler and less prone to technical failure. We also designed an easy-to-use help card for the caregivers as well as a system installation checklist and guide for the interventionists.

We learned the challenges associated with language translation and the need to allocate sufficient resources to this activity. The literacy level of our caregivers was an important design consideration, especially for our Haitian participants. Clarity and pacing of the speech aspects of the system were important, as many of our caregivers were older. Importantly, we learned the value of considering contextual factors when designing home intervention protocols, including consideration of internet access and space constraints within the caregivers' homes. When conducting a home-based assessment, it is important to do an environmental assessment in addition to a user needs assessment (see Chapter 3).

Team meetings were essential for fostering communication among our clinical and technical teams. This enhanced understanding of the project objectives and the various team members' roles and responsibilities. In addition, team meetings enhanced understanding of the needs and characteristics of our caregiver population.

2.6   Conclusions

In this chapter, we discuss the importance of considering the characteristics and needs of user groups in the design process. Our focus was on older adults, but as noted, for most products and systems, user groups extend beyond older adult populations. An important step in the design process is identifying all potential user groups. Sometimes the results of this exercise are surprising as initial thoughts about a user population may be limited. Today, for example, users of technology are likely to include individuals in the later decades. Current and future generations of older adults are different in many dimensions from prior generations. We need to consider them as active users of products and systems.

A second important aspect of design is having an understanding of the needs, characteristics, and preferences of user groups. In this chapter, we stress the importance of individual differences. An important theme of the chapter is the vast heterogeneity of the older adult population. This heterogeneity needs to be considered in the design process as well as in the selection of users for aspects of design such as usability testing. In general, participants should include older adults of varying ages, socio-demographic characteristics, and prior experiences, and with varying functional and health statuses. Involvement of users in design is essential to design success.

References

Czaja, S. J., Boot, W. R., Charness, N., & Rogers, W. A. (2019). *Designing for older adults: Principles and creative human factors approaches* (3rd ed.). CRC Press.

Czaja, S. J., Lee, C. C., Arana, N., Nair, S. N., & Sharit, J. (2014). Use of a telehealth system by older adults with hypertension. *Journal of Telemedicine and Telecare, 20*(4), 184–191. doi:10.1177/1357633x14533889.

Czaja, S. J., Loewenstein, D., Schulz, R., Nair, S. N., & Perdomo, D. (2013). A video-phone intervention for dementia caregivers. *The American Journal of Geriatric Psychiatry, 21*(11), 1071–1081. doi:10.1016/j.jagp.2013.02.019.

Czaja, S. J., Zarcadoolas, C., Vaughon, W., Lee, C. C., Rockoff, M. L., & Levy, J. (2015). The usability of electronic personal health record systems for an underserved adult population. *Human Factors, 57*(3), 491–506. doi:10.1177/0018720814549238.

Lee, C. C., Czaja, S. J., Moxely, J. H., Sharit, J., Boot, W. R., Charness, N., & Rogers, W. A. (2019). Attitudes toward computers across adulthood from 1994 to 2013. *The Gerontologist, 59*(1), 22–33. doi:10.1093/geront/gny081.

Merriam-Webster. (2003). *Merriam-Webster's collegiate dictionary* (11th ed.). Merriam-Webster.

Office of the National Center for Health Information Technology. (2019). *What is an electronic health record (EHR)?* Retrieved from https://www.healthit.gov/faq/what-electronic-health-record-ehr.

Sanger, D. E. (1984, May 17). I.B.M.'s problems with Junior. *The New York Times.*

Taha, J., Czaja, S. J., Sharit, J., & Morrow, D. G. (2013). Factors affecting usage of a personal health record to manage health. *Psychology and Aging, 28*(4), 1124–1139. doi:10.1037/a0033911.

Zarcadoolas, C., Vaughon, W., Czaja, S. J., Levy, J. & Rockoff, M. L. (2013). Consumers' perceptions of patient-accessible electronic medical records. *Journal of Medical Internet Research, 15*(8), e168. doi:10.2196/jmir.2507.

Additional Recommended Readings

Rogers, W. A., Blocker, K. A., & Dupuy, L. (in press). Current and emerging technologies for supporting successful aging. In A. Gutchess & A. Thomas (Eds.), *Handbook of cognitive aging: A life course perspective.* Cambridge University Press.

Schaie, W. K., & Willis, S. (2015). *Handbook of the psychology of aging* (8th ed.). Academic Press.

chapter three

Assessing Needs with Older Adults

At the heart of the design process is meeting the needs of the person who will be using whatever is being designed. Products, environments, and systems (simple or complex) will have end users, and perhaps multiple groups of end users. The Nielsen-Norman Group provides excellent resources for understanding user needs (www.nngroup.com), one example of which is an empathy map. The idea is to understand the many facets of a user – what the person says, does, thinks, and feels. There may be contradictions among these different facets, requiring them to be considered holistically to understand the complexity of the person who will be interacting with the design. Different stakeholders (i.e., user groups) must be considered in the design process – in this chapter we focus on older adults.

3.1   Needs Assessment for Design

From the perspective of technology design, a needs assessment answers the question of "what *should* the technology be designed to do, for whom, and in what context?" Note that it is in opposition to "what *can* the technology be designed to do?" A needs assessment provides guidance and focus for developing solutions to support the needs of the end user, in this case, older adults.

Figure 3.1 shows the design process, beginning with *understand*, which comprises *empathize* and *define*. The "empathize" component focuses on developing an in-depth knowledge of the activity goals of your target users in context (e.g., socio-cultural, physical), as illustrated in the CREATE model (see Chapter 1, Figure 1.1). The "define" component is the integration of the initial efforts to identify gaps that might be filled by a new technology (or by a redesign of an existing technology).

3.2   Needs Assessment Techniques

Needs assessment occurs via multiple methods, including systematic literature reviews, surveys, interviews, observations, participatory design sessions, in-depth case studies, and so on. The goal is to understand the

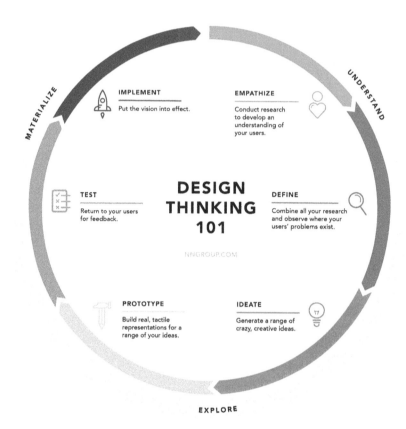

Figure 3.1 Design thinking 101. *Source*: Used with permission (https://nngroup. com/articles/design-thinking/)

needs of individuals (i.e., person assessment), the context of the activities (i.e., environment assessment), and the currently available tools for achieving the person's goals (i.e., comparative analysis).

Consider the following use case: *Carol is 67 and a recently retired high school teacher, living alone in a high-rise apartment. She is a gifted teacher and wants to stay actively engaged with mentoring students and new teachers throughout the United States. She also wants to remain physically active and is especially interested in getting involved in exercise classes where she can learn tai chi. Unfortunately, Carol has progressive macular degeneration and is losing her foveal vision. She does not drive, and her apartment is far away from the nearest bus stop. She has experience with computers from her years as a teacher but has minimal experience with smartphones, apps, tele-video systems, or videogames.*

Use cases provide a valuable starting point to describe the characteristics of the target user population and the context for their activities.

The person assessment can provide some general guidance about technology experience, perceptual capabilities, and activity goals. Interviews with older adults in similar circumstances can provide in-depth information about how these characteristics will influence interactions with technology. Likewise, the context of use in this example is within a high-rise apartment, a single person without technical support, perhaps low bandwidth internet access, and so on. There may be existing apps and software that could help Carol, but an analysis will be required to determine if such tools are accessible to her with her vision challenges and are usable by older adults with minimal experience, if the features meet her needs, and whether adequate instructional support is provided. In sum, the needs assessment should be broad ranging to understand the person(s) being designed for, the activity goals, the environment in which the technology will be used, and currently available technologies, if any, that are attempting to meet the needs.

3.3   Case Study 1: Healthcare Applications

3.3.1   Research Questions

There has been a remarkable proliferation of applications (apps) for smartphones purported to support health self-management activities. To illustrate, a brief trip to the app store (https://apple.com/ios/app-store/) in the fall of 2019 revealed well over 1,000 apps that could be broadly classified as supporting health self-management (see Table 3.1).

The list in Table 3.1 prompts a number of questions. How would an older adult choose one of these apps? How would a family member or healthcare provider know what is best to recommend? Are they designed with older adults' needs in mind? What should designers focus on to ensure usability by older adults?

An initial starting point would be to conduct a comparative analysis of the apps within a category to determine if they are designed with the needs of older adults in mind and would accommodate older adults' motor, perceptual, and cognitive capabilities. Given such large numbers within a category, it might be best to focus on just a few apps within a category. We conducted a comparative analysis of available and popular (i.e., most downloaded) healthcare apps to answer a very general research question: are these apps designed to be usable by older adults? We assumed that such apps have the potential to be useful, given that many older adults manage at least one – often more than one – chronic condition; they are generally motivated to remain as healthy as possible, and there is a trend toward health self-management. Thus, our primary focus was on *usability* to determine if older adults' needs were being met, and, if not, to provide design guidance for elder-focused app design.

Table 3.1 Applications (Apps) for smart phones
related to health self-management

Search term	App store hits (iPhone)
Biking	100+
Blood pressure tracking	100+
Diabetes management	100+
Exercise	100+
Fitness	100+
Mental health	100+
Nutrition	100+
Running	100+
Sleep tracking	100+
Hydration reminder	83
Medication management	57
Doctor locator	49
Mental motivation	29
Health maintenance	17
Online therapy	17
Swimming	4

3.3.2   Illustration of Comparative App Analysis

To examine usability issues across the mHealth domain (Morey et al., 2019), we identified and evaluated one commonly downloaded medication reminder app (Medisafe Pill Reminder [Medisafe Inc.]) and two popular congestive heart failure (CHF) management apps (Heart Failure Health Storylines [HFHS; Self Care Catalysts] and Heart Partner [Novartis Pharmaceuticals Corporation]).

For each type of app, we conducted heuristic evaluations and cognitive walkthroughs to identify where older adults' needs might not be met. These methods are described in more detail in Chapter 4 as standard components of usability research. We mention them here, in the context of needs assessment, because they are also useful tools for identifying unmet needs that might be addressed with new designs.

In the cognitive walkthrough of each app, two different researchers conducted the tasks that users might perform such as data entry or goal-setting, with the goal of identifying where older adults might have difficulty. In the heuristic analyses, we reviewed each app for violations of Nielsen and Molich's (1990) usability heuristics (e.g., consistency of displays, ease of navigation, visibility). We identified similar design issues across the apps that may affect usability for older users including poor navigation, small buttons, low color contrast, small text sizes, data entry difficulties, inadequate data visualizations, and lack of access to help information.

3.3.3   Design Value

The comparative analysis led directly to guidelines for developers of health apps to facilitate use by older adults (see Table 3.2). There were clear themes that emerged across apps that would be challenging for older adults and thereby impede their successful use of the app. This type of analysis provides guidance for the needs of target users that should be translated into new prototypes and ultimately better apps to support their healthcare goals.

3.4   Case Study 2: Digital Home Assistants

3.4.1   Research Questions

Digital home assistants (e.g., Amazon Alexa, Google Home) are becoming popularized as tools that can help people control their environments, engage in social activities, support healthcare activities, and entertain themselves. What are the design features that will make these tools useful to and usable by older adults?

As a starting point we interviewed individuals over the age of 55 who were currently using an Amazon Echo in their home. The goal was to gain an understanding of how they obtained the device, how they set it up, how they were using it, and what additional tasks they might like to perform using the Echo.

3.4.2   Identifying Preferences for Digital Home Assistants

Despite the potential of digital assistants to support older adults, these technologies are not necessarily designed with respect to their unique needs and capabilities. To assess facilitators and barriers to the adoption of these emerging technologies, we explored the attitudes and perceptions of current older adult digital assistant users. We interviewed current users aged 55–91, a convenience sample of early adopters who could provide insights into the story of their adoption, their experiences, and their ideas for additional functionality.

Our script development and our qualitative data analysis were informed by the Unified Theory of Acceptance and Use of Technology (UTAUT2) illustrated in Table 3.3. Following this theoretical framework provided a guide for developing questions that would elicit older adults' preferences and needs for a digital home assistant. Moreover, UTAUT2 provided an organizational structure for the data analysis. Too often,

Table 3.2 Guidelines for designing healthcare app for older adult users

General suggestion	Specific recommendation
Increase button and text sizes	• Design text and buttons to be usable on the smallest device the app will be used on (e.g., make it usable for smartphone users).
	• Use a font size of at least size 30 pt for critical text and at least 20 pt for secondary text. If in doubt, opt for a larger font size. Only use font sizes smaller than 20 pt sparingly.
	• Use large round or square buttons/icons of at least 15 mm in diameter. Avoid narrow/rectangular buttons as these can be difficult to engage with.
	• Increase spacing between buttons/icons to minimize errors with button selection.
Use color effectively	• Keep app themes high in color contrast (e.g., avoid using gray text on white or light-colored backgrounds). Following the WCAG guidelines (World Wide Web Consortium, 2016), we suggest aiming for a text-to-background contrast ratio of at least 4.5:1 for smaller text and at least 3:1 for larger text (at least 18-pt regular font).
	• Opt for bold colors over pale or fluorescent colors.
	• Use color as a tool to categorize different groups of options/icons (e.g., use red to represent all physician or appointment information and blue to represent symptom tracking options).
Ensure within-app consistency	• Keep all button sizes and labels consistent throughout the app.
Simplify within-app navigation	• Use a clear Home Screen that the user can easily navigate back to if they get lost in the app or if they want to use a different function – make sure the home screen is easily identifiable and able to be located with ease from any screen/function (e.g., centered at the bottom of each screen).
	• Limit the number of key functions available from the home screen to 12 or fewer (for tablets) and 6 or fewer (for smartphones). Ensure all key functions are visible on the home screen without the need to scroll.
	• Eliminate scroll bars where possible.
	• Avoid overloading the user with too many options/functions/customizable features – difficult for the user to understand and may lead to incorrect use or failure to use.

(*Continued*)

Table 3.2 (Continued) Guidelines for designing healthcare app for older
adult users

General suggestion	Specific recommendation
Enhance and simplify data visualizations	• Use visualizations where necessary, but keep to a minimum and make as simple as possible. • Avoid overcrowding visualizations with multiple factors or data labels. Present no more than two factors on any one graph/chart, unless completely necessary.
Improve access to help information	• Provide users with a help guide that is easy to identify and access. Ideally, help information should be accessible from the home screen. • Provide step-by-step instructions for all extra options/functions and explain what each option is for/what it does (e.g., Medisafe Pill Reminder's "Medfriend" function). Consider using pop-up hints or a help bubble for the first time a user performs a new task. Direct users to a point of contact (e.g., email address/phone helpline) if requiring additional help.
Improve user privacy and offline access	• Explicitly state what information is private and what information is/can be made public. • Ensure any data sharing with family/friends/physicians is appropriately set up so users do not accidentally share private information. • Allow users to customize what information is shared. Provide explicit step-by-step instructions about how to change privacy/sharing settings. Before users change their sharing settings, include an explicit reminder/confirmation screen asking whether users are sure they wish to increase/decrease their privacy settings. • Clearly define unfamiliar terms/functions (e.g., Medisafe's Medfriend) that may lead to users accidentally sharing personal information. • Ensure users can log on to the app and enter/edit/save data in the absence of an internet connection.

Source: Used with permission, Morey et al. (2019).

interviews are conducted in a very *ad hoc* manner wherein the questions might be leading (e.g., focusing only on what people like or only on what they do not like) or the content covered is not comprehensive (e.g., covers only usefulness or ease of use but not enjoyment or instructional support needs).

The aim was to identify factors that influence the initial adoption, as well as continued use, of digital assistants for adults 55 years of age and older. Thus, the interview script focused on the participants'

Table 3.3 Unified theory of acceptance and use of technology (UTAUT2) model factors and definitions

Model factor	Definition
Social influence	The extent to which users perceive that important others (e.g., family, friends) believe they should use a particular technology
Performance expectancy	The degree to which using a technology will provide benefits to consumers in performing certain activities
Effort expectancy	The degree of ease associated with the use of technology
Facilitating conditions	The user's perceptions of the resources and support available to perform a behavior
Hedonic motivation	The degree to which a user enjoys using the technology
Price value	User's perception of the value of the technology relative to the cost
Habit	The result of prior experiences with the target technology

Source: Venkatesh et al. (2003, 2012). Table used with permission, Koon et al. (2020).

initial reasons for device purchase, their experience with device setup, and how they first learned to use the digital assistant. The majority reported that another individual, most commonly a family member (e.g., son/daughter), purchased and set up the digital assistant for them. Among those who purchased the device for themselves, half relied on assistance from others to set the device up, whereas the other half set it up on their own. Participants who set up the digital assistant themselves reported higher technology experience compared to those who obtained assistance. These data illustrate both the importance of social influence and facilitating conditions for using a new technology. Social influence refers to the extent to which users perceive that important others (e.g., family, friends) believe they should be using a particular technology. In this case, participants reported that others purchased the device for them because they thought they would enjoy it. The support provided for setting it up served as a facilitating condition for use. The digital assistant users we interviewed may not have taken the initial purchasing step on their own, but became consistent users of the devices.

The factors in the UTAUT2 model have primarily been utilized to predict behavioral intention to adopt a technology rather than continued use. However, we found that factors in the model did influence the continued use of the Amazon Echo.

First consider *performance expectancy*. This factor represents the usefulness of a system – does it perform as expected, will it meet the user's needs, does it provide benefits? Participants reported positive considerations that affected their use:

- "It has to be music, the thing I enjoy the most. The sound is just stupendous."
- "I can … talk to her and say … just add this to the list and then it's done."
- "I don't know if there's anything I don't like to do on the Amazon Echo. …"

However, there were also negative comments relating to performance expectancy that influenced the user experience:

- "I guess the thing I least enjoy would be asking her a question that she says, 'I don't know what you mean' … So that's probably my only frustration with her …"
- "I'm limited because … I don't really know about other services it could provide."
- "Well, it doesn't seem necessary to use the Echo and use the email. Or the telephone. I use the telephone a lot."

There were positive and negative comments related to other factors in the model as well. To illustrate briefly, *effort expectancy* relates to the ease of use, with a positive comment being, "It's even easier than using my smartphone … all I have to do is speak" and a negative comment being, "Sometimes when you want the answer to a question you have to word it just so, or she doesn't quite understand you."

Enjoyment using a system will influence continued use; in the UTAUT this factor is referred to as "hedonic motivation." Participants' comments related to aspects they enjoyed about their digital assistant ("I've had a lot of fun with it.") as well as negative experiences ("I feel that it's intrusive.").

In addition, there may be *facilitating conditions* that influence the continued use of a digital assistant. These might include instructional support or assistance from others, namely, resources that support use. Participants mentioned this factor in positive ("So far, everything she's come up with that I've attempted to use has not been an issue.") as well as negative ("I don't really know about other services it could provide.") ways, based on their experience.

3.4.3   Design Value

These data indicated the factors that were important to the continued use of the Echo, facilitating continued use, or imposing a barrier to the continued use of a digital home assistant. The results provide guidance for design iterations of these tools based on the needs and preferences of older adults. One key theme was the importance of providing instructional support. It is common for older adults to receive these devices as gifts – however, that is only the first step. They need to know how to set them up and how to add features that will enhance their value over time.

Another theme was that there were positive aspects that contributed to their continued use (e.g., enjoyment, ability to do a variety of activities) as well as negative aspects that might ultimately lead to discontinued use (e.g., communication frustrations). Lastly, it was clear from the interviews that the older adults, even though they were experienced users, had a very limited understanding of the range of capabilities of the digital assistant. They wanted to do more things with it, but either were unaware of its capabilities or were unable to get it to work to accomplish these additional tasks.

Clearly it is not sufficient to provide functionality for a device – the users need guidance for setting it up and getting it to work, initially. They then need support for learning what the device can do for them and how to interact with it effectively to enhance their continued engagement over time.

3.5   Case Study 3: Design of a Personal Robot for the Home

3.5.1   Research Questions

A project might start with a technology in mind, and then use needs assessment techniques to refine the specific design. For example, there is a general belief in the robot design community that personal robots have a tremendous potential to support the needs of older adults in their homes.

How exactly might a robot fulfill older adults' needs? What activities would older adults be willing to have a personal robot do for them? What characteristics of a personal robot would be acceptable to older adults in their homes? These questions illustrate that a needs assessment should consider specific functions for a technology tool as well as preferences and attitudes that will influence the acceptance and integration of technology into the lives of older adults.

3.5.2   Understanding Older Adults' Needs and Preferences for Robots

We addressed these questions in a series of studies related to the design of personal robots for older adults. First, we conducted a literature review to identify the broad home activity needs of older adults with the goal of understanding where robot support might be helpful (Mitzner et al., 2014). Many adults over 65 years of age require some assistance, yet it is important for their feelings of well-being that the assistance not restrict their autonomy. We evaluated older adults' diverse living situations and the predictors of residential moves to higher levels of care as well as their needs for assistance with activities of daily living (ADLs) or instrumental activities of daily living (IADLs) when living independently or in a long-term care residence.

Most older adults live in private homes with a spouse; however, many live alone (approximately 30% of those over age 65; 39% of those 85 and older). Close to 50% of women aged 75 and over live alone. About 32% of adults aged 65 and older live with other relatives, such as in a multigenerational household. Nearly 11% of adults aged 65 and older live in private homes with informal care, and almost 5% live in private homes with formal care. The percentage of individuals living in long-term care settings of any type increases with age: 1.3% (65–74), 3.8% (75–84), and 15.4% (over age 85).

Robots that assist with the activities listed in Table 3.4 could be of benefit to a large number of older adults living both in private homes and in long-term care settings. Not only do a substantial percentage of older adults require assistance for these everyday activities, if those needs are unmet, but it may result in unwanted moves to environments that provide higher levels of care.

To deploy a robot to provide assistance with these tasks in a person's private home, technical challenges would need to be overcome. The robot

Table 3.4 Robot assistance opportunities for older adults living in private homes vs. long-term care, based on needs assessments

Activity	Private home	Long-term care (assisted living or skilled nursing)
Activities of daily living (ADLs)	Ambulation Bathing/showering Dressing Transfer	Ambulation Bathing/showering Dressing Feeding Transfer Toileting
Instrumental activities of daily living (IADLs)	Errands/shopping Housework Making phone calls Managing medications Preparing meals	Errands/shopping Housework Making phone calls Managing medications Preparing meals
Chronic health conditions	Heart disease Diabetes Cerebral vascular disease Chronic obstructive pulmonary disease (COPD)	Heart disease Diabetes Stroke COPD/emphysema Mental illness (e.g., depression, dementia)

Source: Adapted with permission from Mitzner et al. (2014).

would need to be robust enough to require little maintenance or troubleshooting, especially in the home of an older adult living alone. The robot would need to be able to negotiate the navigation environment of the home such as dynamic and static obstacles, uneven floors, stairs, varying lighting conditions, door thresholds, clutter, and pets. The needs of diverse stakeholders would have to be considered, including informal caregivers, formal caregivers, older adults who require care, as well as remotely located robot operators as these different user groups might all interact with the robot in different ways.

In the immediate future, it may be more economically feasible to deploy robots in long-term care residences than in private homes. A single robot could assist multiple care recipients and care providers. Robots may be able to support older adults sufficiently to enable them to continue residence in assisted living longer before moving to more expensive, skilled nursing residences.

Robots are embodied agents that take up space in an environment and have a particular form factor. To better understand older adults' needs and preferences for a robot in their home, we provided older adults with the opportunity to interact with a personal robot in a simulated home environment, the Aware Home at Georgia Tech. Older adult participants observed a mobile manipulator robot autonomously demonstrating three tasks: delivering medication, learning to turn off a light switch, and organizing home objects. We administered pre- and post-exposure questionnaires about participants' opinions and attitudes toward the robot, as well as a semi-structured interview about each demonstration. The demonstration of the robot did, in fact, influence older adults' acceptance. There was a significant increase, pre vs. post, in positive perceptions of robot usefulness and the ease of use, and the older adults showed greater openness to robot assistance after exposure.

To illustrate the details of the data we obtained, consider the task of delivering medication. When asked, "what is your first impression of the medication delivery task?" most participants responded very positively, although two thought the robot moved too slowly, and one thought it was not currently useful to them but might be in the future. This latter point is worth elaboration. It is not uncommon when older adults are asked about the value of a technology support, they view it as useful for someone else or for themselves in the future, but may not see a current need. Consequently when assessing older adults' perceptions of usefulness, we often encourage them to think counterfactually (if I did need support, then this would be useful) or in terms of others (how might this be helpful for other people your age) as a way to gain insights into their design preferences.

We asked participants to elaborate on why they held certain first impressions. As depicted in Figure 3.2, they reported many factors as

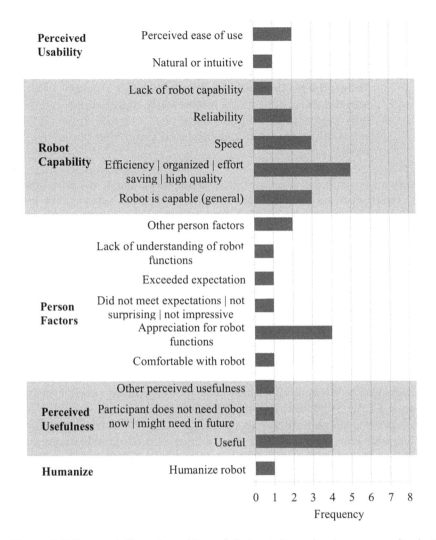

Figure 3.2 Factors influencing older adults' opinions about a personal robot delivering medication. *Source*: Used with permission (Beer et al., 2017)

influencing their first impressions. The reasoning behind their impressions was largely the robots' capability. For example, they recognized that the robot would save them time and effort by retrieving medications. Person-related factors were mainly categorized as a positive appreciation or liking of the robot.

When the older adults were asked whether they would prefer the robot to deliver a bottle versus individual pills, most indicated that they would prefer the bottle. This preference was driven by the older adults' concerns about reliability; namely, there would be less likelihood of error, if the robot delivered the bottle and the person could verify that it was the correct medicine. Preference also depended on the robot's capability as well as their own; one participant stated,

> Today, the bottle would be fine. If the roles change and the robot is thinking more clearly than I about how many [pills] do I take, then yes … ideally [the robot] give you what you need and only what you need.

Clearly a key aspect of human–robot interaction is trust. We investigated this construct in more detail for the context of home support. We interviewed older adults who had experience with care providers (Stuck & Rogers, 2018). Trust is an essential element for older adults and robot care providers to work effectively. As trust is context dependent, we explored what older adults perceived as supporting trust in robot care providers within four common home-care tasks: bathing, transferring, medication assistance, and household tasks. Older adults reported three main dimensions that support trust: professional skills, personal traits, and communication. Sub-themes included ability, reliability, safety, as well as the robot's benevolence (i.e., kindness, goodwill), the material of the robot, and the companionability of the robot.

One aspect of a needs assessment that should not be overlooked is the person's own capabilities and how they influence needs and preferences. In the Stuck and Rogers (2018) study we found a positive relationship between self-efficacy of operating a robot and trust preference. Participants were asked about their confidence in being able to operate a robot. In general, participants had low to neutral self-efficacy in operating the robot; however, there was a positive trend between self-efficacy and trust preference in a robot. Self-efficacy in operating a robot may make older adult users more open to trusting a robot to assist them. If so, robots for use by older adults should be easy to use and training should be provided so the older adult feels confident in getting the robot to perform a task and potentially trusts it more.

3.5.3 Design Value

This series of studies provided valuable information about design needs for robots that older adults will interact with, especially in the context of healthcare. The research findings culminated in a framework to guide robot design to maximize the success of the human–robot interaction for older adults (Rogers & Mitzner, 2017; see Figure 3.3).

Much like the CREATE framework depicted in Figure 1.1 of Chapter 1, this representation of the factors important to robot design is multifaceted

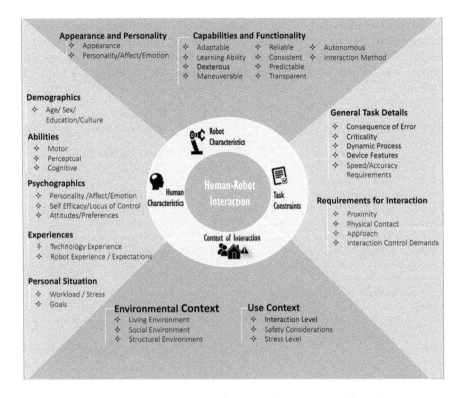

Figure 3.3 Human–robot interaction framework. *Source*: Used with permission (Rogers & Mitzner, 2017)

and contextual. The characteristics of the human have to be considered, as well as the characteristics of the robot itself. Moreover, the details of the task being carried out, as well as the context of the interaction, are important considerations. The specific items listed in the framework emerged in large part from the aforementioned needs assessment studies wherein these issues were identified as relevant.

3.6   Conclusions

A primary theme of the chapter is that designers need to be designing *with* older adults, rather than *for* them. A user-centered design philosophy requires understanding users' needs. However, the concept of "needs" should not be constrained to what is necessary but should also include what is desired. The needs refer to what the technology needs to provide. Awareness of preferences and attitudes will determine acceptance of technologies and thus should not be ignored in the needs assessment phase. The context of use is also an important design consideration. Further, the translation of knowledge about users' needs into defined gaps is a critical aspect of the design process (see Figure 3.1). This translation process leads to specifications for design goals. Systematic needs assessment efforts will lead to technology designs that enhance the quality of life of older adults.

References

Beer, J. M., Prakash, A., Smarr, C-. A., Chen, T. L., Hawkins, K., Nguyen, H., Deyle, T., Mitzner, T. L., Kemp, C. C., & Rogers, W. A. (2017). Older users' acceptance of an assistive robot: Attitudinal changes following brief exposure. *Gerontechnology, 16*(1), 21–36. doi:10.4017/gt.2017.16.1.003.00.

Koon, L. M., Blocker, K. A., McGlynn, S., A., & Rogers, W. A. (2020). Perceptions of digital assistants from early adopters aged 55+. *Ergonomics in Design, 28,* 16–23.

Mitzner, T. L., Chen, T. L., Kemp, C. C., & Rogers, W. A. (2014). Identifying the potential for robotics to assist older adults in different living environments. *International Journal of Social Robotics, 6*(2), 213–227. doi:10.1007/s12369-013-0218-7.

Morey, S. A., Stuck, R. E., Chong, A., Barg-Walkow, L., H., & Rogers, W. A. (2019). Mobile health apps: Improving usability for older adult users. *Ergonomics in Design, 27,* 4–13.

Nielsen, J., & Molich, R. (1990, April). Heuristic evaluation of user interfaces. *CHI '90 Proceedings of the SIGCHI Conference on Human Factors in Computing Systems* (pp. 249–256). doi:10.1145/97243.97281.

Rogers, W. A., & Mitzner, T. L. (2017). Human-robot interaction for older adults. *Encyclopedia of computer science and technology* (2nd ed., pp. 1–11). CRC Press.

Stuck, R. E., & Rogers, W. A. (2018). Older adults' perceptions of supporting factors of trust in a robot care provider. *Journal of Robotics, 2018,* Article ID 6519713, https://doi.org/10.1155/2018/6519713.

Venkatesh, V., Morris, M. G., Davis, G. B., & Davis, F. D. (2003). User acceptance of information technology: Toward a unified view. *MIS Quarterly,* 425–478.

Venkatesh, V., Thong, J. Y., & Xu, X. (2012). Consumer acceptance and use of information technology: Extending the unified theory of acceptance and use of technology. *MIS Quarterly,* 157–178.

World Wide Web Consortium. (2016). *Contrast (minimum): Understanding success criterion 1.4.3.* Retrieved from https://w3.org/TR/UNDERSTANDING-WCAG20/visual-audio-contrast-contrast.html.

Additional Recommended Readings

Harrington, C. N., Mitzner, T. L., & Rogers, W. A. (2015). Understanding the role of technology for meeting the support needs of older adults in the USA with functional limitations. *Gerontechnology, 14*(1), 21–31. doi:10.4017/gt.2015.14.1.004.00.

Norman, D. A. (2013). *The design of everyday things: Revised and expanded edition.* Author.

Preusse, K. C., Gonzalez, E. T., Singleton, J., Mitzner. T. L., & Rogers, W. A. (2016). Understanding the needs of individuals ageing with impairment. *International Journal of Human Factors and Ergonomics, 4*(2), 144–168. doi:10.1504/ijhfe.2016.10003159.

Smarr, C.-A., Mitzner, T. L., Beer, J. M., Prakash, A. Chen, T. L., Kemp, C. C., & Rogers, W. A. (2014). Domestic robots for older adults: Attitudes, preferences, and potential. *International Journal of Social Robotics, 6*(2), 229–247. doi:10.1007/s12369-013-0220-0.

Wooldridge, A. R., & Rogers, W. A. (in press). Understanding the patient, wellness, and caregiving work of older adults. In R. J. Holden, & R. S. Valdez (Eds.), *The patient factor: A handbook on patient ergonomics.* CRC Press.

chapter four

Implementing Usability Methods

Much thought and effort goes into the conceptualization, development, production, and marketing of new products. Nonetheless, users may have difficulty using the product as intended, for a whole host of reasons. The impetus behind usability assessments is to understand the source of these use challenges throughout every stage of the process, so as to improve the design or instructional support and fulfill the potential of the products. Note that we will use the term "product" in this chapter as shorthand; similar methods apply for environments, systems, training programs, websites, etc. – the main ideas are the same. Moreover, the basics of the different usability methods are constant regardless of the user group, but there may be nuances that are specific to older adult users. We will highlight these in the case studies with older adults provided in this chapter.

4.1   Usability Methods

There is no single best way to assess usability. In many cases, different methods will be used at different stages of the design process. Each method provides certain insights that can be informative. Often multiple methods will be needed (often iteratively) to fully understand the usability challenges and how best to address them. The goal of this chapter is to provide a short primer of some of the most common usability methods, along with several case studies that illustrate their application in different contexts.

There are excellent resources that provide more details about each method (see Additional Recommended Readings). In addition, the User Experience Professionals Association (UXPA; founded in 1991 as the Usability Professionals Association) supports the development and dissemination of usability tools and techniques worldwide. The UXPA site (https://uxpa.org/) provides links to books, meetings, short courses, webinars, and other valuable resources. They provide a very clear definition of the user experience. It includes:

> Every aspect of the user's interaction with a product, service, or company that make up the user's perceptions of the whole. User experience design as a discipline is concerned with all the elements that

together make up that interface, including layout, visual design, text, brand, sound, and interaction. UE works to coordinate these elements to allow for the best possible interaction by users.

Another excellent resource is the Nielsen Norman Group (https://nngroup. com/). The site provides open access to articles and videos that cover the range of usability methods and applications. A good place to start is with Usability 101 (https://nngroup.com/articles/usability-101-introduction-to-usability/). Herein it summarizes five major components of usability that should be considered, and measured, in the design process:

- *Learnability*: How easy is it for users to accomplish basic tasks the first time they encounter the design?
- *Efficiency*: Once users have learned the design, how quickly can they perform tasks?
- *Memorability*: When users return to the design after a period of not using it, how easily can they reestablish proficiency?
- *Errors*: How many errors do users make, how severe are these errors, and how easily can they recover from the errors?
- *Satisfaction*: How pleasant is it to use the design?

Usability testing may be misconstrued as only involving tests with users of final product designs. However, usability assessment should be considered much more broadly. Another good source for usability resources is www.usability.gov. There are multiple usability methods that can be implemented at different stages, and not all of them involve direct testing of users. For example, heuristic evaluation and task analysis are typically conducted by human factors experts, and the outcomes are used to guide (re) design efforts. Of course, it is also necessary to directly assess usability with the people who are ultimately going to use the product. These stakeholders may be from different groups and use the product in a variety of ways. As such, the usability testing should include representative users, performing the range of expected tasks, and in the actual (or simulated) use context.

4.1.1   Cognitive Walkthrough

A cognitive walkthrough is a technique that is performed by someone with training and expertise in human factors. The idea is to perform specific tasks on the product in a step-by-step fashion with consideration for the cognitive demands placed on the user at each step. The goal is to consider the different aspects of usability described above (learnability, efficiency, memorability, errors, satisfaction) to identify where users might encounter difficulties. The key is to be as detailed as possible when developing the set of representative tasks that a user might perform. The person doing the analysis should imagine they are performing the tasks as a participant and think about the capabilities and limitations for the participant while performing each task. It is useful to conduct the cognitive walkthrough with different participants in mind, for example, persons with mild cognitive impairment or an older adult with visual limitations. For each activity, points of potential confusion are identified and addressed in the design iterations.

The UserFocus website provides illustrations of a cognitive walkthrough and identifies four key questions (https://userfocus.co.uk/articles/cogwalk.html):

1. Will the customer realistically be trying to do this action?
2. Is the control for the action visible?
3. Is there a strong link between the control and the action?
4. Is feedback appropriate?

These questions are asked at each step of the task, and they are very useful for identifying usability problems. Note that the cognitive walkthrough is not performed by the end user, but it is a preliminary analysis that should be conducted by the design team to identify problems that can be remedied before the product is even tested with the target user group. Chapters 3 and 8 both provide illustrative examples of cognitive walkthroughs.

4.1.2 Heuristic Evaluation

A heuristic evaluation involves reviewing a specific product with consideration for general usability rules of thumb (i.e., heuristics). Consider the design of a website, for example. As a general rule, information should be laid out in a consistent manner across sections of the site, terminology and labels should be used consistently, colors should have consistent meanings, and so on. Sometimes these conventions might be intentionally violated to highlight certain information. But, *in general*, following the heuristic of consistency will yield a more usable site.

Table 4.1 describes ten usability heuristics for user interface design. These heuristics provide an excellent starting point for the analysis of a system in any phase of development from early prototypes to fully functioning products. Other sets of heuristics and guidelines are tailored for specific platforms (e.g., mobile devices, Gómez et al., 2014) or contexts (e.g., speech interfaces, Weinschenk & Barker, 2000).

4.1.3 Task Analyses

A task analysis is a method for decomposing a task into the steps required to perform it. The goal is to specify, in as much detail as possible, what a user needs to do to achieve the task goal. A classic example used in

Table 4.1 Ten usability heuristics for user interface design by Jakob Nielsen

Heuristic	Description
Visibility of system status	The system should always keep users informed about what is going on, through appropriate feedback within a reasonable time.
Match between the system and the real world	The system should speak the users' language, with words, phrases, and concepts familiar to the user, rather than system-oriented terms. Follow real-world conventions, making information appear in a natural and logical order.
User control and freedom	Users often choose system functions by mistake and will need a clearly marked "emergency exit" to leave the unwanted state without having to go through an extended dialogue. Support undo and redo.
Consistency and standards	Users should not have to wonder whether different words, situations, or actions mean the same thing. Follow platform conventions.
Error prevention	Even better than good error messages is a careful design that prevents a problem from occurring in the first place. Either eliminate error-prone conditions or check for them and present users with a confirmation option before they commit to the action.
Recognition rather than recall	Minimize the user's memory load by making objects, actions, and options visible. The user should not have to remember information from one part of the dialogue to another. Instructions for the use of the system should be visible or easily retrievable whenever appropriate.
Flexibility and efficiency of use	Accelerators – unseen by the novice user – may often speed up the interaction for the expert user such that the system can cater to both inexperienced and experienced users. Allow users to tailor frequent actions.
Aesthetic and minimalist design	Dialogues should not contain information which is irrelevant or rarely needed. Every extra unit of information in a dialogue competes with the relevant units of information and diminishes their relative visibility.
Help users recognize, diagnose, and recover from errors	Error messages should be expressed in plain language (no codes), precisely indicate the problem, and constructively suggest a solution.
Help and documentation	Even though it is better if the system can be used without documentation, it may be necessary to provide help and documentation. Any such information should be easy to search, focus on the user's task, list concrete steps to be carried out, and not be too large.

Source: Used with permission, https://nngroup.com/articles/ten-usability-heuristics/

teaching people about task analysis is to have someone explain every step involved in tying their shoes – and have someone else follow the instructions exactly. This demonstration makes clear that even a seemingly simple task has many sub-tasks and that it is quite difficult to articulate a comprehensive task analysis. This task is primarily motor and requires a certain level of dexterity. A task such as using a website may require more cognitive engagement, such as remembering a series of numbers and letters for a password (memory), interpreting an error message (comprehension), or selecting a path of action (decision making).

Once all of the steps have been identified, each step may then be further analyzed in terms of the information the user needs at that step of the process, the likelihood of making an error, and the probable nature of an error. There are different types of task analyses, but, from a usability perspective, the overall goal is the same – to figure out exactly what the user needs to do to successfully use a product. Most task analysis methods require training and expertise to administer (see Adams et al., 2012, 2013, for more details).

4.1.4   User Testing: Formative and Summative Evaluation

There are myriad ways to engage users in the usability evaluation process. A distinction is made between formative evaluation and summative evaluation. In the formative process, users are engaged during the development of the product, often iteratively. An early mock-up of a prototype might be used to illustrate how the product is intended to function, and users can imagine interacting with it and provide insights about parts of it that might be confusing or aspects that seem unnecessarily complex. Then after design adjustments and a working prototype has been developed, user interactions and reactions can be observed to identify continued areas in need of design adjustments.

The term "summative evaluation" generally refers to evaluations of usability for a fully functioning product, perhaps newly on the market or new to a particular user. The summative evaluation might assess initial use as well as longer-term use to understand if usability issues are reduced or perhaps increased over time. For example, some instructions might be required, and once the user has learned the appropriate steps for using a product, the usability difficulties are minimal. Alternatively, as the user becomes familiar with a product, new usability issues might arise (e.g., the system responds too slowly to keep up with practiced user input.)

Regardless of the evaluation phase, several tools are available to evaluate product usability. Measuring completion time and use errors is often helpful for identifying usability difficulties. Patterns of errors across users are especially informative as they indicate a common problem that should be remedied by design (or instruction). An experiment can systematically compare these data for different design formats. Interviews provide insights about details regarding what users found confusing or what they did not like about a product or a specific feature. In addition, there are standard measures that can provide a usability score to help designers gauge the level of usability across people and across products.

One commonly used tool is the System Usability Scale (SUS). After interaction with the product, the users are asked ten questions (e.g., I found the system unnecessarily complex; I felt very confident using the system), with five response options (strongly agree to strongly disagree). The full scale and scoring tool is available from usability.gov. A score of 68 is considered average for this scale; thus if your product scores below 68 then it is not very usable. Designers typically aim to achieve a score of 80 for representative users as a minimum criterion for moving forward with a design. The SUS is not diagnostic in that it does not provide insights into the source of a problem. As such, it is often used in conjunction with other measures such as observations of performance and interviews. Nevertheless, the SUS provides an overall score that imparts a sense of the overall level of usability for your product.

4.2 Case Study 1: Understanding Technology Use Challenges

4.2.1 Research Questions

Interviews are valuable tools for obtaining information about usability challenges. These insights can then be used to guide design – either by redesigning current products or by designing future products. An interview can be conducted with an individual immediately following a product interaction. Another application of this method is to conduct a group interview (sometimes called a "focus group"), and the discussion might revolve around a single product or a class of products. Mitzner et al. (2010) used this latter approach to gain insights into usability challenges that older adults were experiencing, in general, across different types of technology that were becoming prevalent in their everyday lives.

The overarching research goal of this study was to explore the details of older adults' attitudes about technology broadly in terms of the range of technologies they used in different contexts (home, health, or work), their likes, dislikes, use challenges, training needs, etc. A group interview approach was ideal for this goal because it provides an open and exploratory method for collecting qualitative data on technology use, insights into the details of actual usage, as well as perceived advantages and disadvantages of technology in different domains.

4.2.2   Illustration of Group Interview Approach

At first blush, a group interview (or an individual interview) simply requires asking people a series of questions. However, the form, sequence, nature, and content of those questions all require careful consideration. The objectives are to be comprehensive, but not overwhelming; to guide the discussion, but not be leading; and to allow a flow of conversation, but to keep people on the topic. Consequently, the script for the interview needs to be thoughtfully developed and pilot-tested before the study begins.

Figure 4.1 provides some excerpts from the Mitzner et al. (2010) script. It was designed to facilitate discussion about the range of older adults' technology use and their attitudes about technology in the domains of home, work, and health. The script was pilot tested with two groups of older adults ($n = 10$) to ensure that the discussion questions were clear and prompted discussion relevant to the issues of immediate interest. The technology was defined as "electronic or digital products and services." This level of detail illustrates the preparation that precedes a successful group interview study.

The selection of individuals for the groups also needs to be decided systematically. If the groups are too small, the conversation might lag; if too large, everyone might not be able to contribute. If the groups are too similar in makeup, there might not be much diversity of opinion. If they are too different, there may be some discomfort given

Setting the context:

This research is part of a large 5-year grant, funded by the National Institutes of Health...our goal is to better understand technology use by older adults. Your information will help us to conduct research on these topics and, ultimately, to develop technologies that are more useful and easier to use.

First, I'm going to summarize what you will be doing, but I will remind you of these questions during the session. We are interested in understanding what types of technologies you currently use or have ever used in the performance of everyday activities, how you use these technologies and what you think about them.

The activities we will focus on today are {2 circled topics: home / work / health}. What we would like you to talk about is how you have been using technology in your everyday experiences with these activities......

Some rules for the conversation:

We are really interested in your personal experiences and ideas.

If your ideas are different from someone else's: speak up, we are interested in a range of experiences.

You can follow-up on something that another person says as long as you do not interrupt that person.

Speak clearly and loudly enough for everyone to hear and for us to be able to transcribe the recording later.

Please speak one at a time. Remember, we are recording the session, and we cannot understand the tape when more than one person is talking at a time.

Starting the discussion:

First, we would like you to tell us what kinds of technology items you use for healthcare either for yourself or others and how often do you use them. You should think about technology items here in a broad sense and not limit yourself to very technologically advanced items. Also, think about the occasions in which you use technology for healthcare, such as when you communicate with health care professionals, when you learn to use medical devices, or gather information about diseases. {Assistant moderator lists on white board}

"Does anyone have anything to add?" {Limit item brainstorming session to 10 minutes per area.}

For those of you who have used {discuss each technology mentioned in brainstorming} what do you like and dislike about using this technology for health care?

What do you think is the best way to learn how to use _____for healthcare?

Do you think you would like additional help with or instruction with _____ for healthcare? If so, what types of tasks you think you might want help or instruction for?

What type of information or training do you think would help you learn how to do (each task)?

Figure 4.1 Excerpts from the group interview script used in the Mitzner et al. (2010) study.

different viewpoints or experience. Mitzner et al. (2010) included a total of 113 community-dwelling older adults, interviewed in 18 groups, ranging in size from 4 to 9 participants in each group (note that only two domains were discussed in each group to keep the length of the discussion manageable; the order of the discussion was counterbalanced). The group interviews were conducted at three sites: Georgia Institute of Technology, Florida State University, and University of Miami. Both males and females were included, and the overall sample was ethnically diverse. The decision was made to limit some groups to individuals with lower education (less than college) and other groups to individuals with higher education (at least college). Education level is a proxy for socio-economic status, and people with lower education individuals might have had less access to technology. This grouping was presumed to lead to better (i.e., more comfortable for the participants) discussions of technology attitudes.

4.2.3   Design Value

The group interviews yielded a rich data set. An initial conference paper (Mitzner et al., 2008) delved into specific preferences for training across technology and provided insights into how older adults would like to be trained and by whom. The primary Mitzner et al. (2010) paper presented a systematically developed comprehensive coding scheme to understand technology attitudes for a diverse sample of older adults across different domains. From the perspective of usability, the "dislikes" provided the most insight about use challenges that older adults were having – and hence identified potential targets of opportunity for more usable designs. Reasons for disliking a technology were most frequently related to the inconvenience brought on by using the technology (e.g., technology causing interruptions in their lives; requiring or increasing mental or physical effort) or features of the technology (e.g., too many or too few features or programming options; poor content or programming quality; and poor output quality, such as the quality of sound or the picture of a visual display). Specific quotes from the interviews provide insights into the design issues and guidance for potential solutions (e.g., [There are] *too many options* [on the automated telephone menu]; *There are too many choices* [on digital cameras]; [The calculator] *doesn't show enough information*.)

In sum, understanding the usability challenges people experience in the context of everyday use provides guidance for next-generation products and new features for existing products, and may identify needs for instructional supports. The format of the group interview led to an in-depth discussion and a rich, informative data set.

4.3   Case Study 2: As Easy as 1, 2, 3?

4.3.1   Research Questions

Medical devices have moved from the formal healthcare environment into people's everyday lives. That is a positive trend because it often enables patients to track their health conditions better (e.g., blood pressure meter, pulse-oximeters, blood glucose monitors [BGMs]), which may ultimately lead to better long-term health outcomes. Given the importance of the data provided by these technologies, usability issues are particularly important. Rogers et al. (2001) set out to assess the usability of BGMs that were on the market in the early 2000s. Although the design of BGMs has greatly improved in the last decades, this case study illustrates well the value of a task analytic method. The approach can be broadly applied.

One of the most readily available BGMs on the market was advertised as being "as easy as 1, 2, 3 … *simply set up the meter, check the system, and test your blood.*" Yet users were having difficulty using it properly, getting faulty readings, and putting themselves at health risks. The purpose of the Rogers et al. (2001) analysis was to understand why these usability issues were occurring.

4.3.2   Task Analytic Approach

To examine the true complexity of the BGM referred to in the advertisement, the authors of the Rogers et al. (2001) study conducted a detailed task analysis, a portion of which is reproduced in Table 4.2. Using a hierarchical task analysis method, they determined that in reality, there were 52 sub-steps required for a person to use the BGM effectively to test their blood. Setting up the meter required 6 steps, checking the system required 22 steps, and checking one's blood glucose level required 24 steps. At each step, there was certain knowledge required and opportunities for errors. As such, it was not surprising that users were having difficulty using the device properly. (This was not an isolated instance; a task analysis of a different BGM on the market at that time required 61 steps.)

4.3.3   Design Value

Task analyses provide explicit information about the steps necessary to perform a task. For complex tasks, there may be a lot of steps necessary, and it may not be possible to design a simpler system. The design value of the task analysis in that case is information about where individuals are most likely to make an error and what type of error. Designers can provide reminders or other types of environment supports to reduce the task

Table 4.2 Partial listing of the steps required for the tasks involved in using a blood glucose meter (the full list has 52 steps)

Task number	Task	Task/knowledge requirements	Feedback	Potential problems
1.0	Set up the meter			
1.1	Select the display language			
1.1.1	Press and hold the C button	Location of C button	None	Cannot locate button
1.1.2	Press and release the On/Off button	Location of On/Off button	None	Cannot locate button Fails to release button
1.1.3	Release the C button	Location of the C button	None	Cannot locate button
1.2	Code the meter			
1.2.1	Turn on the meter	Location of On/Off button	The meter displays last reading	Cannot locate button
1.2.2	Compare the code numbers on the meter and test strip package	Location of the correct code number	None	Cannot find correct code number on the package
1.2.3	Press the C button until the codes match	Location of the C button	Code changes on the display	Enters incorrect code number
2.0	Check the system			
2.1	Perform a check strip test			
2.1.1	Make sure the test area is clean	Location of the test area	None	Test area not cleaned
2.1.2	Turn the meter on	Location of On/Off button	The meter displays last reading	Cannot locate button
2.1.3	Wait for the meter to say, "INSERT STRIP"	Location of display	The meter displays instructions	Does not observe instructions on display, inserts the strip too early

(Continued)

Table 4.2 (Continued) Partial listing of the steps required for the tasks involved in using a blood glucose meter (the full list has 52 steps)

Task number	Task	Task/knowledge requirements	Feedback	Potential problems
2.1.4	Slide side 1 of the check strip into the test strip holder	Location of the test strip holder, proper orientation of the check strip	None	Insert the check strip incorrectly
2.1.5	Wait for the meter to say, "APPLY SAMPLE"	Location of the display, correct procedure	The meter displays instructions	Does not remove the check strip from the holder, applies blood or control solution, does not wait for instructions
2.1.6	Slide the check strip out of the test strip holder	Correct procedure	None	Does not remove check strip
2.1.7	Wait for the meter to say, "INSERT SIDE 2"	Location of the display, correct procedure	The meter displays instructions	Does not wait for the instructions
2.1.8	Slide side 2 of the check strip into the test strip holder	Location of the test strip holder	None	Inserts check strip incorrectly
2.1.9	Wait for the meter to count down from 4 to 0	Location of the display, correct procedure	The meter displays count	Does not wait for the meter to count down

Source: Used with permission from Rogers et al. (2001).

demands or support the use of cognitive resources to improve the usability of the system (see Figure 4.2). These findings can also be used to guide the design of effective instructional support, as we did in Mykityshyn et al. (2002), which is described in Chapter 7.

Moreover, it is important for users to have accurate expectations about a product. If it is presented as trivially easy to use, they will blame themselves if they make errors, rather than look for instructions or seek help. If, instead, their expectation is that they will need to invest some time and energy to learn to use it, they may be more motivated to do so.

4.4   Case Study 3: Designing Better Ballots

4.4.1   Research Questions

The U.S. presidential election in 2000 was controversial due to concerns about the accuracy of counting certain types of ballots. To assess the usability of different voting methods, Jastrzembski and Charness (2007) compared the performance of younger (aged 18–26) and older adults (aged 64–77) for two voting machine types and two ballot formats. The primary research question was whether the usability differed across methods and across age groups.

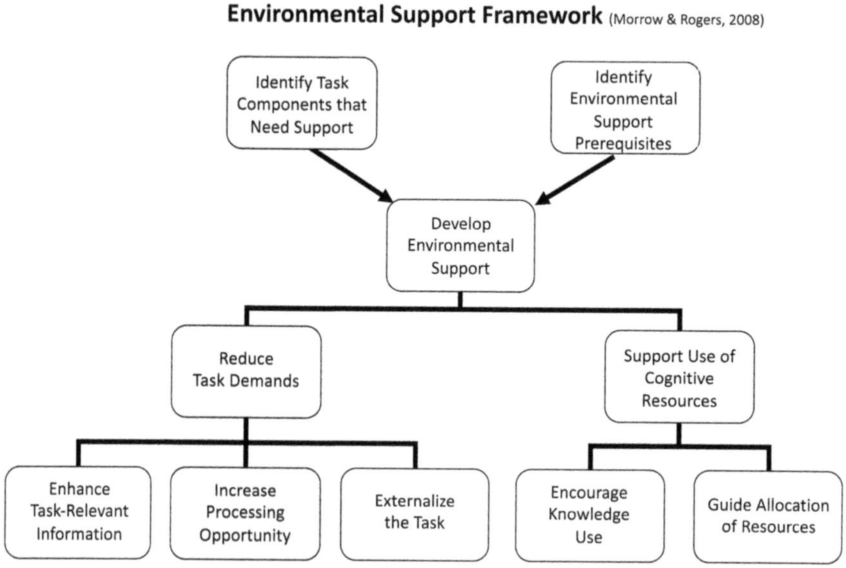

Environmental Support Framework (Morrow & Rogers, 2008)

Figure 4.2 The Environmental Support Framework for design. *Source*: Reprinted with permission from Morrow and Rogers (2008)

4.4.2 Testing Users Experimentally

The voting machines in the study were either touchscreen systems (with a light pen for input) or a mixed design that required typing in the number of the candidate and then using the light pen/touchscreen to approve and cast a vote. The ballot format was either a whole ballot on one screen or a page ballot that separated information into different categories (e.g., governor on one page, secretary of state on another page). These two variables were crossed to create four conditions, depicted in Figures 4.3a and 4.3b.

Participants were given a hard copy list of candidates to vote for and could refer back to this list during the voting task. Each participant was tested using each of the four experimental conditions, counterbalanced. The error and completion time data are presented in Figure 4.4. Compared to younger adults, the older adults made more errors and were slower using the mixed voting machine, regardless of ballot format. They were especially slower for the mixed voting machine with the page ballot (although their errors were slightly lower than those for the whole ballot format).

Jastrzembski and Charness (2007) noted that although the error rates were seemingly low, a 1.9% error rate would have translated into ~2.4 million invalid ballots in the 2004 U.S. presidential election. In addition, older adults made nearly three times as many errors as younger adults in the worst case, which would make their voting disproportionately invalid.

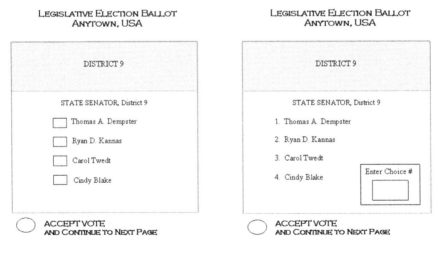

Figure 4.3a Ballot options for the page ballot. On the left is the touchscreen option wherein the candidate selection and vote acceptance are both done with the touchscreen. On the right is the mixed option that requires first typing in the number and then using the touchscreen to accept the vote. *Source*: Used with permission (Jastrzembski & Charness, 2007)

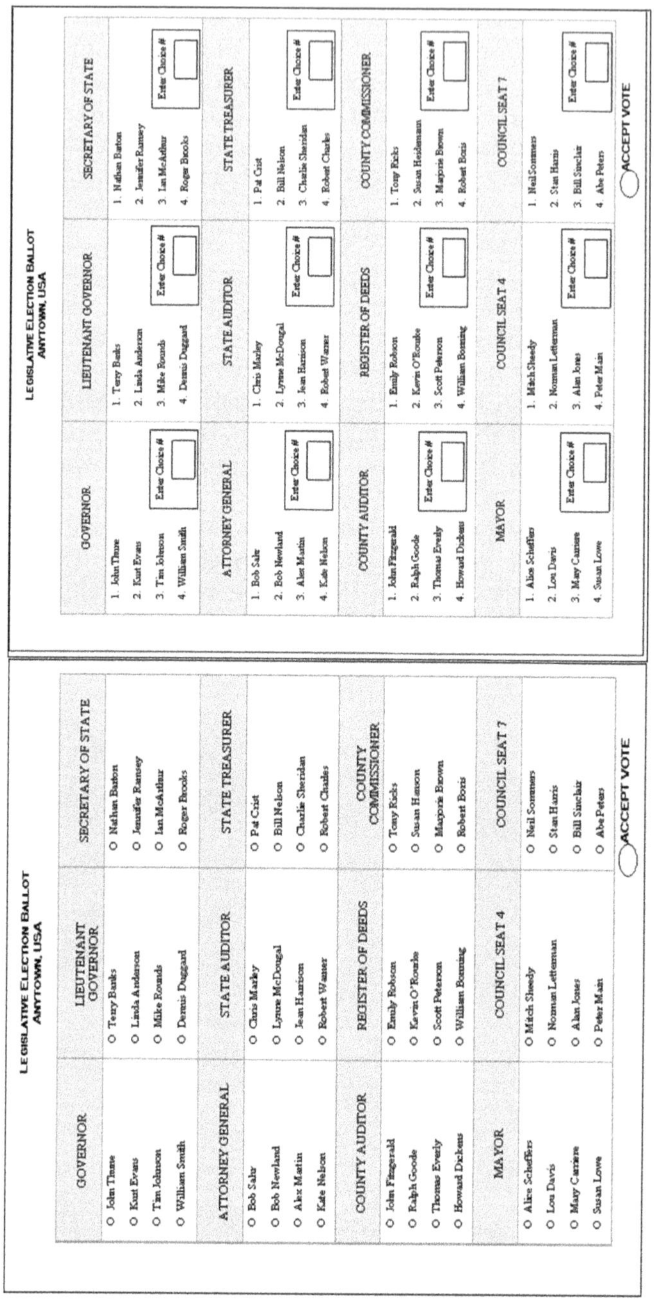

Figure 4.3b Ballot options for the whole ballot. On the left is the touchscreen option wherein the candidate selection and vote acceptance are both done with the touchscreen. On the right is the mixed option that requires first typing in the number for each selection and then using the touchscreen to accept the vote. *Source:* Used with permission (Jastrzembski & Charness, 2007)

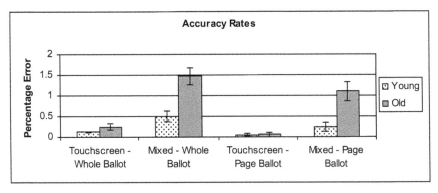

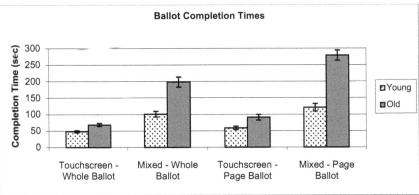

Figure 4.4 Age-related differences in accuracy (top, represented as error rates) and completion time (bottom) for different ballot formats (see text for descriptions). *Source*: Used with permission (Jastrzembski & Charness, 2007)

4.4.3 Design Value

One general finding from Jastrzembski and Charness (2007) was that the touchscreens were faster for both age groups. Generally, not requiring users to switch between typing and touch input is more efficient, and this was clearly illustrated in their data. Moreover, errors were lower for the touchscreen, making this the overall preferred voting machine. But what if, for cost or other reasons, the community had no choice but to use the mixed voting machine – should the ballot format be presented as a whole or page by page?

As is often that case, there can be tradeoffs between speed of performance and accuracy. In this study, the page ballot was slower but more accurate. The decision-makers would have to evaluate the cost of more errors against the cost of people having to wait longer in line. This type of relative cost analysis is necessary to guide design choices with consideration for broader socio-political contexts. Additional complexities with voting machines are cybersecurity controls (which could interact with ease of use).

Other points illustrated in this study are first that there were age-related differences in the impact of the designs on usability. Second, there was an interaction between the machine and the ballot format. Thus the format of a display must be considered in the context of the technology platform being used. The mantra of usability testing: representative tasks performed by representative users in representative contexts.

4.5   Case Study 4: Usability of Fitness Applications (Apps)

4.5.1   Research Questions

There has been a proliferation of fitness apps that are designed to help people with their wellness goals. The apps often track movement (usually steps) and diet (food intake) and provide feedback to users about their own performance. Some devices allow for goal setting and for social comparison to others. Some activity trackers are non-wearable logs (e.g., MyFitnessPal.com) requiring minimal input from the user, whereas others are wearable activity trackers that also include logs (e.g., Fitbit One), which do require user input. This emerging class of technology has the potential to help older adults with health self-management, self-efficacy, and healthy habits. Despite this potential, many older adults do not use activity trackers for their health-tracking needs. Older adults may not use activity trackers because of usability barriers, which leads to several research questions that were addressed in a multimethod study conducted by Preusse et al. (2017). First, are fitness trackers usable, in general; do they adhere to usability principles? Second, are older adults able to use the fitness trackers, do they like them and find them easy to use and useful?

4.5.2   Insights from Multiple Usability Methods Guide Design and Deployment

Preusse, Mitzner, Fausset, and Rogers (2017) examined the potential barriers and facilitators to activity tracker acceptance for older adults in a multimethod, two-phase study. In Phase 1, they conducted a heuristic analysis of two activity trackers to assess overall usability that could elucidate barriers related to the perceived ease of use and perceived usefulness, which predict technology acceptance. Phase 2 was a field study wherein 16 older adults (aged 65–75) were assigned one of the trackers to use for four weeks; the field study included usage data, questionnaires, diaries, and interviews that allowed for informed user feedback (both negative and positive) to emerge and provided insight into acceptance over time.

The summary of the Preusse et al. (2017) heuristic evaluation is presented in Table 4.3. Note that some issues are related to the overall category of these types of devices and were not specific to the particular product being evaluated. Moreover, there were instances of design decisions that violated multiple heuristics. The evaluation revealed potential barriers to acceptance in the areas of Consistency and Standards, Visibility of System Status, and Error Prevention. To illustrate (a) Consistency: MyFitnessPal.com allowed exercises to be entered with a dropdown menu on some, but not all, of the exercise entry pages, which could make it difficult for users to know how to enter data; (b) Visibility of System Status: it was difficult for users to easily know the sensitivity setting of the Fitbit One device, which could cause inaccurate automatic data logging; (c) Error Prevention: the delete function for the food log of MyFitnessPal.com was unlabeled and made the format for adding/deleting foods challenging. These heuristic violations provide valuable insight into the redesign of these products to enhance their usability.

The field trial yielded a wealth of data about attitudes before, during, and after interactions with the fitness apps. Positive attitudes provide avenues for designers to include in future products or to emphasize in their advertisements. Negative attitudes provide insights for improvements in design or training. We focus here on the negative only because that is often the goal of usability testing – to uncover what aspects of a product can be improved. Figure 4.5 shows the coded data from the final interview. Overall, inaccuracies coupled with difficult formats (e.g., layout, how information is added), particularly in entering data on food logs, were the main barriers to acceptance. These categories, augmented by quotes from the interviews as well as diary entries, provide clear guidance to designers about the usability challenges that could be addressed in future design iterations or in new products.

4.5.3   Design Value

Often multiple usability methods will be utilized to gain insights into usability challenges. The different methods provide windows through which to view the design of a product. The heuristic analysis is a systematic examination conducted by experts, to identify potential usability challenges. Perhaps these usability issues can be remedied in a redesign, but, if not, the insights provide guidance for target instructions to support users and minimize their frustrations and likelihood of making use errors.

Assessing usability challenges directly with target users is invaluable. In the Preusse et al. (2017) study, assessments were made before,

Table 4.3 Heuristic evaluations of Myfitnesspal.com and the Fitbit One

Heuristic	Violation	Activity tracker	Example	Potential user difficulty
Consistency and standards	Used the same coloring of the website's features for advertisements or premium service links	Fitbit.com	• Some small blue links for logging additional measurements (e.g., weight) opened up additional free boxes for users to type in data. • Other small blue links for adding a new measurement (e.g., glucose) did not allow the user to enter in more boxes and rather prompted the user to purchase a premium service.	• May make it difficult for users to discriminate services that are part of the activity tracker and those that are not. • May cause users to end up on pages they did not want to be on.
		MyFitnessPal.com	• Some advertisements along the banners of the website were in similar colors to those of Myfitnesspal.com (orange and blue). • Links to "Related Ads" were in the themed colors of orange and blue.	
	Color meaning changed across graphs	Fitbit.com	• Progress graphs on the dashboard did not use the same colors consistently on all graphs.	• Having to learn new associations may make reading multiple graphs more challenging.
	Dropdown data entry menus were inconsistently available	MyFitnessPal.com	• On some log pages such as EXERCISE\database a dropdown menu was provided to add exercise. • On other log pages, such as EXERCISE\ExerciseDiary\AddExercise, the user's only option was to start typing in an exercise to search.	• May make it difficult for users to know how to enter data, as they must learn different methods for each page.

(Continued)

Table 4.3 (Continued) Heuristic evaluations of Myfitnesspal.com and the Fitbit One

Heuristic	Violation	Activity tracker	Example	Potential user difficulty
Consistency and standards and visibility of the system status	Inconsistent navigation bars	MyFitnessPal.com	• The typical navigation bar used throughout the rest of the user's account was replaced by a novel and unique navigation bar upon clicking on the "MyBlog" page.	• Novel navigation bars are an unnecessary, second navigation system that the user must learn to use efficiently and may be confused with the primary navigation system.
		Fitbit.com	• On Fitbit.com, when the user went to the help page, the navigation bar completely disappears.	• The elimination of a consistent navigation can make it difficult to know where a user is on the website, if a user is still logged in, and how to get back to the user's logs.
Visibility of the system status	Battery life of device was not visible on the device	Fitbit One (device)	• The user could only view battery life of the device either when it was plugged into the computer or when the user was on the website. • The user could not view battery life levels on the technology otherwise (e.g., when on-person).	• Could make it difficult for new users to know when to recharge.

(Continued)

Table 4.3 (Continued) Heuristic evaluations of Myfitnesspal.com and the Fitbit One

Heuristic	Violation	Activity tracker	Example	Potential user difficulty
Visibility of the system status and error prevention	Device sensitivity setting, which impacts log accuracy, was not obviously communicated	Fitbit.com and Fitbit One (device)	• Inaccuracies of the device's automatic step and sleep logging could have been improved by adjusting sensitivity settings. Sensitivity settings were not prominent because they were not viewable on device or across the website (e.g., the heading). • Sensitive settings were only viewable online under Settings/Device. • Users were not introduced to the sensitivity setting to check accuracy during their first few steps.	• Users may not be aware if the device is in a certain sensitive state and that adjusting that state could prevent errors in automatic step and sleep tracking.
Error prevention	Unlabeled delete icons	MyFitnessPal.com	• On several logs, to remove an entry a user could hit the delete key, which was a small red circle with a horizontal white line. Although multiple delete symbols may exist in the user's knowledge (e.g., x's and trash cans), the small red circle was not labeled. • No confirmation message appeared upon clicking the icon to ask the user if he or she was certain that the entry should be deleted.	• Users may accidently delete logs by not knowing what the icon does, especially when learning how to use the technology.

Source: Used with permission from Preusse et al. (2017).

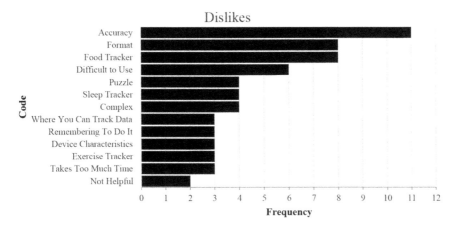

Figure 4.5 Coded interview data illustrating themes that older adults did not like during the eight weeks of using fitness apps. *Source*: Used with permission from Preusse et al. (2017).

during, and after interactions with the product and comparisons could be made across two different types of products. We focused our summary here on the usability challenges and dislikes because these provide direction for design improvements. However, it is important to also emphasize the positives of the study. The older adults liked many aspects of both fitness apps. With minimal instructions they were able to use the apps. They used the apps across the four weeks of the study, and the majority of them were positive about their intentions to use them even after the study was over. Thus designers should never assume that older adults will not be interested in using their products. Older adults are not "technology-averse"; just like people of all ages, older adults want to use products that are useful to them and usable by them.

Collectively, these data provide valuable insights into the design of products that meet the needs and capabilities of older adults. Health apps have a tremendous potential to support health self-management by older adults, and including them in usability assessments will increase the likelihood that such technology will be adopted by older adults as part of their daily health regimens.

4.6 Conclusions

A primary theme of the chapter is that usability assessment is a process – different methods provide different information and may be most relevant at different stages of the product design. The usability methods require training to be most effective. It is not sufficient to just ask users what they think – the interview script must be carefully developed and the answers coded in an unbiased manner. Observing users is useful, but the behavioral observations must be interpreted in a systematic and objective manner. Reviewing a product for potential user challenges is necessary but is much more valuable if it follows well-established heuristics known to relate to usability. Fortunately, there are resources available to help designers develop the skills to implement these usability methods. Formative and summative evaluations are invaluable for the design process and increase the likelihood that the resultant products will be usable and adopted by older adults.

References

Adams, A. E., Rogers, W. A., & Fisk, A. D. (2012). A guiding tool for task analysis methodology. *Ergonomics in Design, 20*, 4–10.

Adams, M. E., Rogers, W. A., & Fisk, A. D. (2013). Skill components of functional task analysis. *Instructional Science, 41*(6), 1009–1046. doi:10.1007/s11251-013-9270-9.

Gómez, R. Y., Caballero, D. C., & Sevillano, J. L. (2014). Heuristic evaluation on mobile interfaces: A new checklist. *The Scientific World Journal, 2014*, 1–19. doi:10.1155/2014/434326.

Jastrzembski, T. S., & Charness, N. (2007). What older adults can teach us about designing better ballots. *Ergonomics in Design, Fall: The Quarterly of Human Factors Applications, 15*(4), 6–11. doi:10.1518/106480407x255198.

Mitzner, T. L., Boron, J. B., Fausset, C. B., Adams, A. E., Charness, N., Czaja, S. J., Dijkstra, K., Fisk, A. D., Rogers, W. A., & Sharit, J. (2010). Older adults talk technology: Their usage and attitudes. *Computers in Human Behavior, 26*(6), 1710–1721. doi:10.1016/j.chb.2010.06.020.

Mitzner, T. L., Fausset, C. B., Boron, J. B., Adams, A. E., Dijkstra, K., Lee, C. C., Rogers, W. A., & Fisk, A. D. (2008). Older adults' training preferences for learning to use technology. *Proceedings of the Human Factors and Ergonomics Society 52nd Annual Meeting* (pp. 2047–2051). Santa Monica, CA: Human Factors and Ergonomics Society.

Morrow, D. G., & Rogers, W. A. (2008). Environmental support: An integrative framework. *Human Factors, 50*(4), 589–613.

Mykityshyn, A. L., Fisk, A. D., & Rogers, W. A. (2002). Learning to use a home medical device: Mediating age-related differences with training. *Human Factors, 44*(3), 354–364. doi:10.1518/0018720024497727.

Preusse, K. C., Mitzner, T. L., Fausset, C. B., & Rogers, W. A. (2017). Older adults' acceptance of activity trackers. *Journal of Applied Gerontology, 36*(2), 127–155. doi:10.1177/0733464815624151.

Rogers, W. A., Mykityshyn, A. L., Campbell, R. H., & Fisk, A. D. (2001). Analysis of a "simple" medical device. *Ergonomics in Design, 9*(1), 6–14. doi:10.1177/106480460100900103.

Weinschenk, S., & Barker, D. (2000). *Designing effective speech interfaces.* Wiley.

Additional Recommended Readings

Czaja, S. J., Zarcadoolas, C., Vaughon, W., Lee, C. C., Rockoff, M., & Levy, J. (2015). The usability of electronic personal health record system for an underserved adult population. *Human Factors, 57*(3), 491–506. doi:10.1177/0018720814549238.

Geisen, E., & Bergstrom, J. R. (2017). *Usability testing for survey research*. Elsevier.

Nielsen, J. (1993). *Usability engineering*. Academic Press.

chapter five

Simulation for Design

In the field of human factors, simulation can serve a variety of important purposes. For example, driving simulators can help assess a driver's fitness to drive under dangerous roadway conditions, and flight simulators can help train pilots to perform complicated flight maneuvers. Critically, these goals can be accomplished without putting the driver or pilot at risk for a crash that might harm themselves or others. Simulation can also be a valuable *design* tool. This chapter explores how simulation can aid the design process, introduces relevant concepts and issues to consider when using simulation for design, and presents two case studies in which simulation methods were used by CREATE investigators to understand, and also solve, design challenges involving older adults.

5.1 Simulation Value for Design

The goal of many designers is to improve users' interactions with complex and dynamic systems. Sometimes these systems can be difficult for designers to access, modify, and test to determine whether the implementation of design changes can improve user experience, reduce errors, or increase efficiency. In other cases, systems are safety critical, and studying performance and design changes using these systems can put individuals and those around them at risk for serious injury or death. For example, it would be dangerous to investigate whether certain changes to an air traffic control display might benefit performance by first implementing these changes at airports. Further, when considering design solutions, different types of performance errors may be relatively rare and related to low-frequency, idiosyncratic events. Rare performance errors under atypical conditions are difficult to study in the field, making the process of determining whether design changes can reduce these errors a challenge. Certain types of dangerous motor vehicle crashes, for example, are thankfully relatively infrequent. Finally, in some cases, designers face the prospect of trying to maximize the design of systems that are still under development (e.g., new autonomous vehicle features). In all of these situations, well-designed simulation studies offer a host of benefits to the design process.

5.1.1   Simulation Overview

In a very broad sense, simulation is the imitation of a process or event. In this chapter, *simulation* refers to a means to test users' interactions with a target system without involving the system itself. In some simulations, computer programs are used to create a virtual environment or interface with which the user interacts. This is typically the case with driving simulators and flight simulators. However, simulation does not need to involve computer-generated environments or even electronic components. For example, in the case of mannequins used to simulate cardiopulmonary resuscitation (CPR), users are interacting primarily with a physical object rather than a virtual environment, though more advanced simulators may have embedded electronics to measure performance and provide feedback. Of primary importance, as much as possible, the simulation should look and behave like the target system of interest with respect to all factors that could influence performance and should accurately capture the user's performance as they interact with the simulated system.

5.1.2   Simulation Use in Design

Although simulation can be useful for a number of purposes, including performance assessment and skills training, this chapter focuses on simulation for design. All of the aforementioned design challenges can be addressed through a well-designed simulation. First, systems that are difficult to access and manipulate (e.g., displays within the control room of a nuclear power plant) can be simulated in the laboratory, and system changes can often be implemented more easily in simulated rather than actual systems and environments. Second, as in the case of driving (and the case of nuclear power plant design), testing design changes in the field can put system users and those around them at risk. If a countermeasure expected to reduce traffic crashes is ineffective or even worse, unexpectedly increases rather than decreases crash risk, drivers may be seriously injured or killed in the evaluation of this countermeasure. Simulation allows for the evaluation of the design of safety-critical systems without these risks. In addition to harm to people, some errors may cause damage to the system itself and result in costly production delays and repairs. Sometimes serious errors result from users encountering rare events in the field. Rare events in simulations can be programmed to occur more frequently, allowing for a more efficient evaluation of designs to prevent errors associated with unexpected but significant events. Given their rarity of some performance errors, simulations can also be difficult to determine whether design changes made a difference. However, simulations can often provide more subtle performance metrics indicative of participants *almost* making this error or being at risk for committing this error (e.g., "near misses" in terms of traffic crashes). Finally, simulation can be used to aid in the design of systems that do not exist yet. Simulated prototype systems, for example, can be implemented in driving simulator studies of novel autonomous vehicle functions.

5.1.3   Domains of Application

Simulators and simulations have been used in a variety of domains to benefit the design of systems for older adults. One of the most notable examples is transportation-related simulations (e.g., driving, flight). Driving simulators range from relatively low-fidelity desktop personal computer (PC)-based simulators to high-fidelity, immersive simulators with a 360-degree view of the driving environment and motion base platform to emulate the forces of acceleration and deceleration. Due to age-related changes, older adults are more likely to be seriously injured or killed when a crash occurs, and changes to perception and cognition put older adults at greater risk of being involved in certain types of crashes. Simulation allows for the safe testing of new and existing roadway countermeasures to reduce older adults' crash risk. One example is discussed later in this chapter as a case study involving investigating countermeasures to reduce wrong-way driving. As mentioned previously, although driving simulators can be used to understand crash risk and the effectiveness of crash countermeasures, they can also serve as important driver training and assessment tools.

In addition to advanced 3D simulations involving an immersive virtual environment, many other everyday tasks can be simulated with the assistance of specialized computer software. Simulations can provide fine-grained performance measures while older adults engage in a variety of instrumental activities of daily living (IADLs), including managing finances and healthcare through technology. For example, the CREATE team used PC-based automatic teller machine (ATM) simulators to better understand the types of errors early older adult ATM users made and the effectiveness of different methods of ATM training in preventing these errors. CREATE has also used simulation to discover usability challenges older adults face when using patient portals – websites and apps that allow older adults to access and interpret their own health records. Developed patient portal simulations were based on popular portals currently being used. Across many different domains, simulation can identify design challenges of systems that support instrumental activities of daily living and design solutions to alleviate these challenges. Once problems have been identified, stimulated solutions can be implemented, and the design of these systems can then be tested again.

5.2   Simulation Implementation

Although simulations can be beneficial to the design process, their design
and implementation can be complex, and many decisions need to be made
regarding how simulations look and feel, the types of data to collect, and
how often data should be collected. This next section provides important
information to consider and guidelines for how simulations should be
implemented to aid in designing for older adults, followed by case studies
following these guidelines.

5.2.1   Capturing the Behavior of Interest

In the use of simulation for design, careful consideration needs to be
given to how to best capture the behavior or behaviors of interest. Often
designers focus on errors and task completion times. Simulation allows
for these types of measures to be easily recorded. However, in many
simulations the measurement of more subtle behaviors can provide
additional valuable information. For example, in driving simulation, in
addition to being able to assess whether or not participants experience
a crash in the simulated environment, the distance between the participants' own simulated vehicle and other vehicles in the simulated
environment can provide a continuous measure of crash risk. A participant can have a number of near misses in a simulated driving task,
demonstrating extremely dangerous driving behaviors, yet still experience no simulated crashes. Many computer-based simulations allow for
the state of the system and user inputs to be recorded continuously with
a high degree of temporal and spatial precision. In the development of
simulations, designers should consider possible performance measures
within this data stream that demonstrate the risk for making an error or
that are indicative of user uncertainty or confusion (e.g., the prolonged
time between the completion of one step and the next). In the absence
of overt errors, these measures provide important insights into system
usability.

Realism is another important factor to consider in achieving the goal
of capturing the behavior of interest. *Realism* (also called *fidelity*) is the
degree to which a simulation looks and feels like the real thing. If the
simulated system does not look, sound, and feel like the target system,
it may not engage the same perceptual and cognitive mechanisms, and
thus performance within the simulated system may not reflect target
system performance. Higher realism is typically preferred. However,
perfect or near-perfect fidelity may not be necessary for many research
questions, and higher fidelity simulations also tend to be costlier to
implement. Designers should consider the benefits and costs of higher
fidelity simulation. For each research question, attention should be paid

to which aspects of the simulation are important for performance and should be rendered in high fidelity, and which aspects are relatively unimportant can be rendered in lower fidelity. For example, for a driving simulator study that involves daytime driving scenarios, a complex and realistic lighting model may be relatively less important compared to nighttime scenarios; all objects should be clearly visible and well-lit during daytime drives. However, when simulating nighttime drives, a complex and realistic lighting model is important because lighting from headlights and other ambient sources determines whether and when crash reduction countermeasures are visible to the driver. If lighting models are not realistic, countermeasures in the simulated roadway may be visible closer or further than they would be outside of the simulator, possibly resulting in the effectiveness of countermeasures being substantially misjudged. Pilot testing and validation studies, when possible, can help determine whether performance is similar within and outside of the simulation.

Sampling rate (how often data are recorded) is another important factor to consider when capturing the behavior of interest. A sampling rate that is too low may result in important behaviors being missed. At the opposite end of the spectrum, some simulators may output the state of the simulation every few milliseconds. This output allows for an extremely detailed view of behavior and performance, but this may come at the cost of difficult-to-analyze data that must be reduced before any meaningful analysis can take place, as well as the cost of disk space needed to store and back up large files.

5.2.2 Challenges of Using Simulation

Although there are a number of benefits to using simulation during design, there are a number of challenges as well. The previous section discussed some general issues to consider when using simulation across all age groups, but some issues disproportionately impact older adults. Unless these issues are carefully addressed in the design of simulation studies, these studies may lack validity. In the case of simulator or cyber sickness, these studies may also risk the discomfort of older participants to a greater degree compared to younger participants.

5.2.3 Simulator and Cyber Sickness

One of the biggest challenges involving simulations can be *simulator sickness* (also sometimes called *cyber sickness*). Simulator sickness is a risk when study participants must interact with or navigate 3D virtual environments, especially for prolonged periods of time. Examples include the use of driving simulators, flight simulators, and head-mounted virtual reality (VR) systems (e.g., HTC Vive or Oculus Rift). This effect can be particularly true for fixed-based driving and flight simulators in which the simulator itself does not move, but the simulation depicts rapid motion. Symptoms of simulator sickness can include nausea, dizziness, headache, fatigue, and, in more severe cases, vomiting. Although there are a number of competing theories to explain the phenomenon, simulator sickness is most often triggered when there is a conflict between motion information received through vision and through the body (vestibular system). In many simulators, visual information will convey motion, while senses from the body provide information that the body is at rest. For reasons that are not entirely understood, older adults, and especially older women, are at higher risk for simulator sickness.

Simulator sickness can be reduced by designing simulator scenarios that minimize the mismatch between visual and vestibular information. During driving simulation (particularly with fixed-based simulators), making turns in the simulated vehicle can create a large discrepancy between visual information and expected vestibular forces. The same is true for driving scenarios that involve acceleration and deceleration (e.g., stopping for a red traffic signal, and accelerating once the signal turns green). When possible, these maneuvers should be avoided. Unfortunately, many maneuvers that are risky for older drivers in terms of safety involve turns, particularly left turns. Minimizing the number of these maneuvers and the duration of the driving scenario is recommended, but this reduction may come at the cost of the study providing less information.

When working with older adults, especially, it is important to design protocols that consider simulator sickness if it can reasonably be expected. Part of the protocol must involve educating study participants about the symptoms of simulator sickness and instructing them to alert the experimenter as soon as any symptoms are detected so the study can be discontinued. Some participants will try to "push through" symptoms of simulator sickness in order to complete the experiment. This behavior should explicitly be discouraged as continuing the study past the point of the initial simulator sickness will result in substantially greater symptoms. Additional preparations to consider are having water and small snacks available for participants if they feel ill, and in the case of severe simulator sickness, cleaning supplies in the laboratory in the event that participants experience vomiting (which is rare, but should still be prepared for). A participant who experiences severe sickness should be encouraged to wait, rest, and recover before leaving the lab and driving home. A useful tool for identifying simulator sickness to either discontinue the study or use as a way to detect data that may have been compromised by simulator sickness is the Simulator Sickness Questionnaire (SSQ; Kennedy, Lane, Berbaum, & Lilienthal, 1993).

5.2.4   Training and Adaptation

In many instances, technology (e.g., computer, tablet, VR headset) is used to simulate a system being studied. Because older adults often have less experience or proficiency with technology, their simulation performance may be dissimilar to their performance in the field as well as to that of younger adults. For example, an ATM might be simulated with a computer using a computer mouse or touchscreen to make system inputs, even though navigating the real ATM might involve pressing physical and mechanical buttons. To maximize the chance that difficulties experienced during the simulation are related to the design of the system being investigated and not the technology simulating the system, participants should have the necessary technology proficiency to begin with, or this proficiency should be trained before the study starts. An adaptation period can increase the chance that performance within the simulation matches what would be expected outside of the simulation. For example, in a driving simulation, the simulated vehicle may behave differently compared to the participants' own car, and it may take time for participants to adapt to traveling through virtual driving environments (e.g., judging speed and distance in virtual environments). An adaptation and practice drive involving maneuvers similar to those that participants will experience later can help ensure that driving behaviors are naturalistic during the study.

5.2.5   Other Technical Challenges

As mentioned previously, simulation datasets can be large and complex, presenting challenges in terms of analysis and storage. Data sampling rate should be carefully considered in a way that balances data completeness and complexity. However, in general, it is often better to record more information than less information. Large files can be down-sampled later using data processing scripts for further analysis. Additionally, data storage is now relatively inexpensive and is getting cheaper every year. Computational challenges should be considered, as well. Rendering realistic 3D environments in real-time while recording detailed behavioral and performance data can require serious computing power. There is often a tradeoff between processing power and fidelity, as rendering environments in very high fidelity can be computationally expensive. Here too, however, processing power is continuously getting better and cheaper. Realistic virtual reality systems were sold for thousands of dollars just a decade ago, but today more powerful systems can be purchased for just a few hundred dollars off-the-shelf.

Having discussed the benefits of simulation for design, the challenges, and how these challenges can be addressed, we now turn to two specific

CREATE examples in which simulation has been useful in designing for older adults. One example involves simulation to reduce crash risk, and the other involves understanding the challenges older adults with cognitive impairment face when using technology to support IADLs.

5.3 Case Study 1: Wrong-Way Crashes

5.3.1 Research Questions

Wrong-way crashes often occur when a driver intending to access a highway mistakes an exit ramp off the highway for an entrance ramp onto the highway. This error results in an extremely dangerous situation: a vehicle moving at highway speeds going against the flow of traffic. In instances when this vehicle collides with a vehicle moving in the correct direction, the crash forces produced are tremendous and often fatal. Although this type of crash is rare, it is one of the deadliest types of crashes that can occur. Many wrong-way crashes are associated with driver intoxication (drug or alcohol) and occur late at night. However, in many countries, older drivers (especially older drivers experiencing cognitive impairment) are overrepresented as wrong-way drivers in these crashes, with these crashes typically occurring during the day.

The Florida Department of Transportation approached us to investigate this issue using simulation. Simulation afforded the opportunity to study this issue in the laboratory without having to worry about the possibility of dangerous wrong-way entries on the roadway, and under carefully controlled conditions (Boot et al., 2015). We conceptualized wrong-way driving as the result of failing to notice and correctly interpret the various cues surrounding highway exit ramps that distinguish them from entrance ramps. Specifically, the research question asked in this project was whether additional, more visible cues placed around an exit ramp might reduce confusion and deter wrong-way entrances of impaired and older drivers.

5.3.2 Simulation Design

A mid-fidelity driving simulator was used to simulate the task of getting onto a highway from an arterial road (see Figure 5.1). This simulator consisted of three large plasma televisions covering approximately 120 degrees of forward visual angle, a smaller LED monitor positioned below that served as the instrument panel, an accelerator and brake pedal set, a steering wheel, and the driver seat of a car. With the help of software developers and 3D modelers, the driving simulator scenario was modeled after the characteristics of an interchange in Florida associated with a high risk of wrong-way driving. Two driving scenarios were created that

Figure 5.1 Photograph of the mid-fidelity driving simulator used in Case Study 1.

were identical except for the number and nature of wrong-way counter-measures in and around the entrance and exit ramps of the highway. The "standard" scenario featured signs and lane markings recommended by the Manual of Uniform Traffic Control Devices, which in the United States specifies national standards for signs, pavement markers, signals, and other traffic control devices. The standard scenario included exit ramps marked with one "Wrong Way" and one "Do Not Enter" sign. The exit ramps within the "enhanced" scenario featured larger, lower-mounted signs, including multiple "Wrong Way" and "Do Not Enter" signs, and on the arterial road, pavement markers depicted the name of the highway in conjunction with forward arrows to convey to drivers the highway entrance was further down the road. This scenario also featured a "No Left Turn" sign in advance of the highway exit (see Figure 5.2).

5.3.3   Simulation Implementation

In our experience, some older adults encounter initial difficulty adjusting to the dynamics and controls of the driving simulator. Because of this challenge, we had participants first complete an unrelated practice driving scenario before completing the wrong-way task. The goal of this short practice scenario was to allow participants to adapt to the dynamics of the simulated vehicle before the scenarios of interest. The practice scenario increased the likelihood that natural driving behaviors would be observed during test scenarios.

Our implementation of the wrong-way scenario had to consider the fact that older adults are more likely to experience simulator sickness

Figure 5.2 The simulated interchange used in Case Study 1 to explore the effectiveness of wrong-way crash countermeasures. Standard countermeasures (top) as well as enhanced countermeasures (bottom) are depicted to discourage drivers from making a left onto the exit ramp ahead.

compared to younger adults. Unfortunately, one of the most sickness-inducing driving maneuvers in a driving simulator is turning, and it is of primary interest to the current research question: do participants correctly turn onto the entrance ramp or erroneously turn onto the exit ramp? To address this issue, we had participants engage in only two scenarios in which they entered the highway: one with standard countermeasures and one with enhanced countermeasures. While more observations are often ideal, a balance needs to be struck between the benefits of more data and the risk of simulator sickness. An important aspect of the protocol involved informing participants of the risk of simulator sickness as well as providing instructions that at the earliest detection of symptoms they should discontinue the experiment and that doing so would result in no penalty.

With respect to the consideration of measures, one obvious outcome measure is whether drivers turned onto the entrance ramp or exit ramp. A problem with this measure, however, is that wrong-way entrances are rare events. This aspect of the problem led to the consideration of more subtle measures indicative of driver confusion. In addition to measuring wrong-way entries, we also examined driver speed in advance of the exit ramp (which participants had to pass before getting to the entrance ramp). We reasoned that if drivers were confused and were considering taking the exit ramp, they would slow down in anticipation of turning, even if they eventually did not turn onto the exit ramp.

Younger and older drivers were recruited for the study. All had a valid driver's license. To be sensitive to the context of when these crashes occur for different types of drivers, younger adults drove scenarios that depicted nighttime driving, and older drivers drove daytime scenarios. Many younger wrong-way drivers are intoxicated. To simulate this, half of the younger adults recruited were assigned to a simulated intoxication condition in which their vision was distorted with special glasses, and their cognitive resources were depleted with a secondary task. Older adults in these crashes typically are not intoxicated, and older participants did not undergo simulated intoxication. All participants completed the two scenarios (standard, enhanced) in counterbalanced order.

5.3.4 Design Value

This simulation, in conjunction with other laboratory studies, was quite valuable in helping to justify the design decision to introduce a greater number of, and more salient, countermeasures in and around exit ramps. Of the 120 drivers who participated in the study, four participants made wrong-way entries. Two wrong-way drivers were in the group of younger adults who experienced simulated impairment, and two were older drivers. All wrong-way entries occurred in the standard scenario, suggesting a benefit to better-designed interchanges with more salient and redundant cues. However, it is difficult to conclude that the enhanced exit ramps had a statistically significant effect with so few wrong-way entries. A benefit of driving simulation is the ability to observe more nuanced driver behavior. We predicted that confusion regarding whether an exit ramp was an entrance ramp would be evidence in vehicle speed. Even for participants who did not make wrong-way entries, participants passed the exit ramp at significantly higher speeds in the enhanced counter-measure scenario compared to the standard scenario (a difference of 5 miles per hour), suggesting more certainty that the exit ramp was not an appropriate location to access the highway. No interactions were observed with age, suggesting younger and older drivers benefited equally from enhanced countermeasures.

Results suggested a clear benefit to the redesign of signs and pavement markers in and around exit ramps. The use of simulation in this case had several advantages over other potential methods. First, compared to a field study that measures wrong-way entries before and after the implementation of additional countermeasures, driving simulator experiments are in a better position to allow causal conclusions to be drawn. In field studies, factors other than the implementation of countermeasures can cause wrong-way driving rates to fluctuate (e.g., seasonal changes, which may bring more or fewer tourists unfamiliar with the interchanges in the area). In the reported simulator study, scenarios were exactly the same except for differences in countermeasures placed in and around ramps. Because of this degree of control, changes in driving behavior can be more confidently linked to changes in interchange design. Further, if this experiment were instead conducted on the road, participants prone to wrong-way entries might have been placed in real risk of a crash if the new countermeasures were ineffective. Finally, in terms of expense, roadway changes can be quite expensive (cost of not just new materials, but their installation, and the costs associate with traffic control needed to keep roadways clear and motorists safe while countermeasures are modified or installed). In many cases, driving simulator studies can be conducted more cheaply compared to field evaluations. However, depending on the complexity of the modeled driving situation, software development and model generation can be costly.

5.4   Case Study 2: Instrumental Activities of Daily Living Simulation

5.4.1   Research Questions

CREATE has used simulations in other contexts to understand the challenges some older adults face with respect to system interfaces that support the performance of IADLs. Understanding these challenges provides crucial information that can inform the design and redesign of these systems, and the design of training to support the use of these systems. Recently, CREATE developed a series of computer-based IADL tasks to simulate interfaces used to complete banking and prescription refill tasks (Czaja et al., 2017). Research questions related to whether ecologically representative daily activity simulations can inform our understanding of the performance challenges some older adults face, especially older adults with mild cognitive impairment (MCI). MCI represents a degree of impairment that goes beyond typical cognitive aging but is not as severe as dementia. As the population ages, the number of individuals with MCI will increase substantially, as will the importance of designing for older adults with a wide range of cognitive abilities. By understanding performance challenges while older adults accomplish important every-day tasks, designers can consider design alternatives to make these tasks related to independence easier for older adults with and without cognitive impairment. Simulation can play an important role in this research and design process.

5.4.2   Simulation Design

Computer software was developed to simulate interfaces with which par-ticipants interacted. Simulated interfaces were computer-generated and presented on a touchscreen display. As finance and medication manage-ment are two important aspects of remaining independent with advanc-ing age, these factors were the focus of two of the tasks chosen to simulate. One simulation involved asking users to perform tasks with a simulated ATM (Figure 5.3). Another simulation asked users to refill a prescription using an automated telephone menu system.

To capture behaviors as they would occur outside of the laboratory, it was important to consider simulation fidelity. The ATM interface and menu structure were adapted from an ATM system currently used in

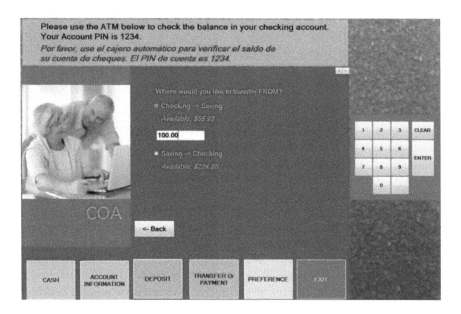

Figure 5.3 Simulated automatic teller machine (ATM) used in Case Study 2.

many locations in the United States. The prescription refill simulation was modeled on one used by a popular pharmacy chain. By modeling tasks closely after existing systems, the generalizability and applicability of study results were enhanced. Further, to account for the linguistic diversity of the older adult population in the United States, simulations were developed in both English and Spanish.

5.4.3 Simulation Implementation

To explore performance in the simulations, older adults with and without MCI were recruited to participate in a study. In the laboratory, they completed simulated IADL tasks as well as traditional measures of cognitive ability that assessed memory, visuospatial processing, and attentional control. For example, for the ATM task participants were asked to complete tasks such as checking their account balances, withdrawing money, and transferring money between different accounts. For the prescription drug task, participants had to refill different prescriptions and schedule a time to pick them up. Although task errors are important with respect to understanding performance (did the user refill the prescription correctly or not?), task completion time, as well as task efficiency (a combined measure that considers both speed and accuracy), was also used to measure performance. A user might use a poorly designed system without making any errors, but the poor design may have wasted time. As simulations were implemented on a touchscreen computer, participants were trained before the study and practiced using the touchscreen. This procedure is similar to the previous case study in which participants were given experience with the simulator first to encourage more natural interactions with the simulation.

5.4.4 Design Value

This exercise in simulation proved valuable in informing design for older adults with a wide range of cognitive abilities. Importantly, it demonstrated that the design of common ATM and prescription refill tasks was challenging for older adults with MCI and that individuals with MCI who had lower cognitive abilities experienced even greater challenges. Forty percent of participants with MCI had difficulties using the simulated ATM, and 13% experienced difficulty performing the prescription refill task. Given these high rates of poor performance, it is likely that older adults with MCI would benefit from better-designed systems. Although not done in this case, the simulated tasks could relatively easily be modified and tested again to determine whether certain design changes might reduce the difficulty of older adults with cognitive impairment.

5.5   Conclusions

This chapter reviewed the potential of simulation to inform design and issues related to using simulation to improve the design of systems specifically for older adults. Among other reasons, simulation is useful when assessing the impact of design changes to safety-critical systems and systems in which errors can result in costly delays or damage to the system. Further, simulation is useful in the study of response to low-probability events and infrequent errors. Assessing performance within simulated systems can provide insight into errors, and simulated design changes can be implemented and tested quickly and efficiently. A number of issues should be considered in general when using simulation to inform design, including issues surrounding simulation fidelity, how to capture the behaviors of interest, and how to deal with the complexity of data derived from simulation. However, two specific issues are especially relevant to older adults with respect to using simulation to study design. First, within simulated 3D environments, older adults – especially older women – are more likely to experience simulator sickness. This is an issue that requires special attention when working with older adults, especially in the context of driving simulation. Second, many simulations are implemented through technology systems with which older adults may have less familiarity or less proficiency. In order to have confidence that difficulties are related to system design and not to interactions with the technology running the simulation, care needs to be taken in terms of participant training and/or selection. Case studies presented here illustrated how simulations could identify system challenges to target for redesign and evaluate the effectiveness of design changes with respect to improving performance. These examples highlighted important issues requiring consideration, how these issues were addressed, and how simulation benefited design.

References

Boot, W. R., Charness, N., Mitchum, A., Roque, N., Stothart, C., & Barajas, K. (2015). *Final report: Driving simulator studies of the effectiveness of countermeasures to prevent wrong-way crashes* (Technical Report BDV30 -977- 10). Florida Department of Transportation. Retrieved from https://fdotwww.blob.core. windows.net/sitefinity/docs/default-source/research/reports/fdot-bdv30-977-10-rpt.pdf.

Czaja, S. J., Loewenstein, D. A., Sabbag, S. A., Curiel, R. E., Crocco, E., & Harvey, P. D. (2017). A novel method for direct assessment of everyday competence among older adults. *Journal of Alzheimer's Disease, 57*(4), 1229–1238. doi:10.3233/jad-161183.

Kennedy, R. S., Lane, N. E., Berbaum, K. S., & Lilienthal, M. G. (1993). Simulator sickness questionnaire: An enhanced method for quantifying simulator sickness. *The International Journal of Aviation Psychology, 3*(3), 203–220. doi:10.1207/s15327108ijap0303_3.

Additional Recommended Readings

Fisher, D. L., Rizzo, M., Caird, J., & Lee, J. D. (2011). *Handbook of driving simulation for engineering, medicine, and psychology*. CRC Press.

Morrow, D., Wickens, C., Rantanen, E., Chang, D., & Marcus, J. (2008). Designing external aids that support older pilots' communication. *The International Journal of Aviation Psychology*, *18*(2), 167–182. doi:10.1080/10508410801926772.

chapter six

Modeling Older Adult Performance

A user-centered design approach is crucial when designing successful products and technologies for older adults. However, this approach can have significant drawbacks in terms of the time and expenses involved, as this process often involves having participants come into the lab for testing. Performance modeling can supplement – and in some cases replace – this process. The goals for this chapter are to show how modeling approaches originated and how they can be used to predict older adults' movement time (Fitts' Law), and to predict performance with different tasks (GOMS modeling for cellphones, driver dilemma zones) by using cases from CREATE research.

6.1 History of Modeling

Modeling has a long history in the physical sciences and for many years relied on mathematics, more specifically, equations, as the modeling language. Physics, for instance, gave us equations to predict movement and force (e.g., $F = M*A$) or energy ($E = M*c^2$). By obtaining and plugging in numeric values on the right-hand side of the equation, one could predict the value for the variable on the left.

The field of experimental psychology is relatively young, with scientific approaches to measuring human performance first being developed in laboratories in Europe in the late 1800s. Investigators such as Ernst Weber and Gustav Fechner started the field of human psychophysics. Wilhelm Wundt, often considered the founder of experimental psychology, was investigating human sensations, perceptions, and attention by conducting systematic experiments on laboratory participants. Hermann Ebbinghaus was outlining principles of human memory by conducting memory experiments on himself. These early psychological scientists also attempted to uncover equations to predict human behavior (e.g., Weber–Fechner Laws relating changes in a stimulus to changes in perception).

Developing theories in psychology were soon applied to practical problems, particularly for military purposes during World Wars I and II in the service of improving human performance on the battlefield and postwar in the factory. Improving human efficiency was the central goal, and the newly developing field of human factors and ergonomics was soon seen as a valuable discipline. Accurately predicting what would be the best design for a piece of equipment or system to optimize human performance drove the attempt to quantify human behavior. Early successes were achieved in predicting human movement time, for instance, through the development of Fitts' Law (1954). At the same time, efforts were also underway to minimize human errors, such as crashes associated with poorly positioned instruments and control devices in the increasingly complex cockpits in military aircraft.

Predicting time and errors is still an ongoing area of study and was responsible in part for the development of cognitive psychology in the 1950s and 1960s. Estimating the time to carry out a task was seen as a critical tool for uncovering the hidden mental processes that fit between the onset of a stimulus and the person's response to that stimulus. When the personal computer revolution took place in the 1970s and 1980s and ordinary people became the target user group, there was a need to design easier-to-use interfaces than the traditional method of issuing commands by typing them. The field of human–computer interaction arose (Card, Moran, & Newell, 1983) by making use of a set of principles and parameters (e.g., the model human processor) to predict performance on complex cognitive, perceptual, and psychomotor tasks. Human factors specialists were quickly welcomed into the nascent computer industry, and the first usability labs were established.

Usability testing (also known as *user testing*) soon became the gold standard for ensuring that a product or service worked well. However, sometimes the price of gold puts such testing out of the range of a finite design budget and/or project development deadlines. Modeling human performance can sometimes fill the gap in time and money budgets.

When it comes to predicting human performance, we are generally interested in two primary features: time and errors. How long does it take to complete a task with Tool A versus Tool B? How likely is it that the task will be completed correctly? The former question is concerned with efficiency in task performance, which in the case of a retired older adult might not be as critical as it is for a paid older worker. The latter question is usually important for assessing product safety, particularly when the cost of errors is high (e.g., medication adherence, setting radiation levels for radiation therapy).

Nonetheless, other features of performance such as how demanding the task is (workload) and enjoyment when using the tool or process (aesthetics/hedonics) are also potentially important features to assess with usability testing because they influence product acceptability. Typically, simulation models are used to predict time and errors.

6.2   Tools That Can Be Used in the Modeling Process: Fitts' Law and GOMS Modeling

We now discuss one relatively simple approach to modeling – using Fitts' Law for predicting movement time – and then a more general systems analysis method: Goals, Operators, Methods, Selection rules (GOMS) modeling, which we have used to help choose among different design options. Generally, our goal is to design for aging users, so we implement such modeling with parameters that have been shown to fit older adults. We begin our discussion with Fitts' Law and provide a case study from CREATE research.

A common design dilemma, particularly for crowded displays such as aircraft cockpits with multiple instrument displays and switches, was where to position controls so that the operator/pilot could respond rapidly to a flight task. Having a control, like a switch, near to the joystick that the pilot used to control flight characteristics of the aircraft would likely produce a faster response than having the control farther away. Making the control large, hence an easy target, should also help, but at the expense of crowding out other controls. How do these two factors jointly predict time to respond? Paul Fitts was one of the first to quantify how a target's size and distance from a start point affected the time to access the target. He observed that movement time to the target was well predicted by what he called the index of difficulty (ID). The two critical parameters for difficulty were *amplitude* (A) of the movement (how far) and *size* (W) of the target (how wide). The equation he generated from data from a reciprocal tapping task became known as Fitts' Law. The critical equation for *movement time* (MT) was:

$$MT = a + b * \log_2 (2A/W)$$

where ID was $\log_2(2A/W)$.

Later, Alan Welford found better empirical fits to data on movement time with a modified version of the equation:

$$ID = \log_2 (A/W + 0.5)$$

When the best-fitting line for the equation passes through the origin (zero point), the *y*-intercept parameter becomes $a = 0$, making the equation simplify to:

$$MT = b * ID$$

As might be expected, given general slowing in information processing with aging, the parameter b interpreted as a processing rate of ms/bit increases with age. Estimates of the value of b as a function of age were initially made by Welford (1977) and later through a type of meta-analysis for parameter values (Jastrzembski & Charness, 2007). Slowing is striking for this form of motor response processing: a typical 20-year-old might be expected to take 100 ms/bit processed, whereas the average 65-year-old takes 75% longer: 175 ms/bit. We can see how this basic equation set can be used to make predictions for interface design in a study by Rogers, Fisk, Collins McLaughlin, and Pak (2005).

6.3 Case Study 1: Using Fitts' Law to Predict Touchscreen Input

6.3.1 Research Questions

The goal for this study, conducted at a time when computing devices had become common at work and were in about half of U.S. households, was to understand how to select the best input device for different types of tasks as a function of the type of task to be performed as well as the characteristics of the user, here user age.

Important for our purposes in this chapter, experiment 2 compared the performances of younger (aged 19–23) and older (aged 51–70) adults using just a touchscreen on tasks such as selecting unstacked and stacked buttons varying in size (five values ranging from 11 mm to 21 mm) and position (3 × 3 matrix of column and row positions), and doing horizontal and vertical scrolling using different-sized scroll bars. Consistent with Fitts' Law, more difficult tasks (selecting a smaller button, scrolling with a longer scroll bar) took more movement time for both younger and older adults, and older adults took more time than younger adults. There were a few interactions observed.

Fitts' Law index of difficulty parameters were estimated for the button selection task, by regressing movement time on the index of difficulty for each task. The regression equation explained between 6% and 14% of the variance for movement time, with the weakest relationship found in the more variable older adult sample. Nonetheless, the slope (B) parameters were in the range of expected values and varied with task difficulty (stacked button positions are harder to discriminate than spaced buttons; see Table 6.1).

The average coefficient values across the two tasks very closely approximate the values (100, 175) estimated a few years later for younger and older adults in the Jastrzembski and Charness (2007) meta-analysis.

6.3.2 How Diversity Influences Performance

This study indicated that different types of devices, tasks, and people (younger, older) had strong influences on performance. One cannot

Table 6.1 Slope coefficient (*B*) for the index of difficulty in younger and older adults when moving to stacked and spaced buttons with a touchscreen

Age group	Younger	Older
B for stacked buttons	112	203
B for spaced buttons	86	146
Mean *B* value	**99**	**174**

simply say that a rotary encoder (indirect positioning device that translates circular movement to linear movement) is superior to a touchscreen (direct) device or vice versa, nor that all people perform equivalently with each device. In fact, the interaction of device and task provided valuable information about when one would want to recommend the use of one device versus the other.

Even so, the good news is that some constants – universals of human behavior – were observed, particularly Fitts' Law slope parameters. The index of difficulty is a powerful predictor of the time to complete a task. Further, knowing the approximate value of the slope constant for different age groups enables one to predict movement time for that age group with reasonable accuracy. The bad news is that there was greater variability in performance in the older adult sample, suggesting that generalizing findings is more difficult for that age group.

6.3.3   Design Value for Modeling

Having a model as well as good estimates for older and younger adult parameters therein may be enough to generate quick (and dirty, because of high variability in older adult performance) estimates for the design process. Fitts' Law is a very useful tool in that respect. From Table 6.1 we can see that stacked buttons selectively disadvantage older users, increasing the index of difficulty by 80% compared to 70% for spaced buttons. Also, it is quite feasible to design an interface by predicting movement time to interface elements under different configurations of target size (*W*) and distance (*A*) from a given cursor position.

6.4   Case Study 2: Using GOMS to Predict Time and Error for a Mobile Device

The Goals, Operators, Methods, Selection rules (GOMS) modeling approach was an important piece of the Card, Moran, and Newell (1983) book, *The Psychology of Human-Computer Interaction*. The goal of such modeling was to assess how different designs would affect human–computer interaction based on the newly developed field of information processing psychology. Information processing psychology proposed that people, when engaged in cognitive tasks, could be viewed as information processing systems analogous to newly developed computer systems. The characteristics and parameters for processing constituted the Model Human Processor (MHP), an idealized human being. An MHP was a collection of memories (e.g., short-term, long-term, iconic) and processes (perceptual, cognitive, motor) that operated according to principles such as bounded rationality, Fitts' Law for movement, and power-law learning. This system had parameters that described capacities and decay rates for memories, cycle times for processor sub-systems, and other processes. The parameters were estimated from experimental studies. However, most of the experiments used young adults (usually college undergraduates) as the sampling population, so such parameter estimates might not adequately describe older adults' performance.

About 20 years later sufficient experimental data had accumulated in the aging literature to provide estimates of these same parameters for older adults. Table 6.2 provides estimates from Jastrzembski and Charness (2007; see also Verhaeghen, 2014). Numbers in parentheses for older adults represent upper and lower bounds for the parameters similar to a confidence interval for a statistical parameter. Numbers in parentheses for the younger adults correspond to the "fast man" and "slow man" estimates in Card, Moran, and Newell (1983).

GOMS modeling applies to *routine task performance*, that is, practiced performance. It would not be applicable to novice performance involving problem-solving activity. Such modeling should provide good estimates of stable performance such as that found in job settings. It is worth keeping in mind that the older adult population shows high inter-individual

Table 6.2 Parameter estimates for younger and older adults

Operator	Younger adult estimate	Older adult estimate
Duration of eye fixation (E)	230 ms (70–700)	267 ms (218–314)
Cognitive processor cycle time (C)	70 ms (25–170)	118 ms (87–147)
Perceptual processor cycle time (P)	100 ms (50–200)	178 ms (141–215)
Motor processor cycle time (M)	70 ms (30–100)	146 ms (114–182)
Fitts' Law slope constant (F)	100 ms/bit (70–120)	175 ms/bit (93–264)
Decay half-life of visual image store	200 ms (90–1000)	159 ms (95–212)
Power Law of practice constant	0.4 (0.2–0.6)	0.49 (0.39–0.59)
Effective capacity of working memory	7 items (5–9)	5.4 items (4.9–5.9)
Pure capacity of working memory	2.5 items (2.0–4.1)	2.3 items (1.9–2.6)

Note: Unit of time is milliseconds (ms).

Source: From Jastrzembski and Charness (2007).

variability; that is, older adults are more unalike each other than younger adults. An even more challenging aspect of older adults' behavior is that they also show higher intra-individual variability compared to younger adults: they are more variable across occasions (show less stability in performance). So, any particular model can be expected to provide ordinal information about design decisions: whether design A is superior to design B and the actual time (or accuracy) advantage for a given design may vary depending on the specific older adult population targeted. Below we provide an example of estimating the time to enter data on a mobile phone.

6.4.1   Research Questions

We have only recently moved away from so-called feature phones to smartphones. For those without experience with feature phones, texting operations required striking a given key on the phone keypad, with a delay less than the timeout period, multiple times to reach one of the three or four letters represented at a given key location in order to register the target keypress. The same is true for entering text information using wired telephones. For instance, the key with the number 2 contains the letter A, B, C, so when entering text, striking the key once would enter A, but if you strike it quickly twice, B is registered, and striking it quickly three times yields C appearing at the cursor position. Timeout periods typically range from 0.5 s to 1.5 s. Pressing the key slowly two times in a row might yield an A and then a second A, instead of a single B. In Jastrzembski and Charness (2007), predictions about task completion time for younger and older users were made concerning the relative efficiency of two phones: a Nokia and a Motorola phone with different keypad layouts and different menu structures for dialing and for texting, using the GOMS model and the parameters shown in Table 6.2. The primary research question under investigation was which mobile phone was better for older users for dialing a number and sending a text message.

An example of instantiating the model is shown for dialing the first three digits of a nine-digit number in Table 6.3 for the Nokia phone, assuming that the thumb is initially hovering over (homed on) the "2 key" when holding the phone one-handed.

As the published article illustrates, the predictions matched extremely well to observed keypresses. Further, knowledge about the physical layout of the keypads on the two phones and their menu structures yielded the predicted advantage for the Motorola phone on the dial task (shorter inter-key distances led to faster movement times) and the Nokia on the texting task (menu structure was easier to navigate, reducing the number of operations to generate a text message). Importantly, GOMS accurately predicted the absolute time to carry out the task components (e.g., steps seen in Table 6.3).

Table 6.3 GOMS modeling of a feature phone dialing task

	Operator	Time (young)	Time (old)	Cumulative time (young)	Cumulative time (old)
ASSUMPTIONS: In default mode for the Nokia – the user holding a cell phone in the preferred hand, dialing with the preferred thumb.					
GOAL: Dial number (268-413-0734) and send a call.					
METHOD: Press numbers and press the green call button.					
Step 1: Fixate the first chunk of numbers on paper (first three numbers)	E	230	267	230	267
Step 2: Encode the first three digits	3 x C	3 (70)	3(118)	440	621
Step 3: Fixate the keypad	E	230	267	670	888
Step 4: Decode the first chunk	C	70	118	740	1006
Step 5: Fixate the first digit	E	230	267	970	1273
Step 6: Dial the first digit	M	70	146	1040	1419
Step 7: Fixate the second digit	E	230	267	1270	1686
Step 8: Dial the second digit	M	70	146	1424.8	1980
	F	84.8	148		
Step 9: Fixate the third digit	E	230	267	1654.8	2247
Step 10: Dial the third digit	M	70	146	1824.8	2568
	F	100	175		

Note: Activities in gray represent observable physical operations. Abbreviations in the Operator column are given in Table 6.2, and values shown in the Time (young) and Time (old) were taken from Table 6.2. Fitts' Law was used to predict the time to move the thumb to the target position from its prior position.

6.4.2 How Diversity Influences Performance

Again, we see that it is necessary to consider user characteristics when designing products. Older adults are usually slower than younger adults, though the gap depends on the specific operator being activated. For eye fixations, there is only about a 15% difference, but for motor processor cycles older adults take twice as long. So, interfaces that rely on eye tracking might be expected to show much smaller age differences than ones that require a mix of operators with relatively few eye fixations. Similarly, age-related slowing that varies by type of process predicts that providing environmental support (Morrow & Rogers, 2008) – putting information out in the world that can be fixated, rather than having to be stored in memory and retrieved – ought to benefit older users differentially because eye fixation slowing ratios with age (1.15) are much smaller than cognitive operator slowing ratios (1.69; see Table 6.2).

6.4.3 Design Value

Because GOMS modeling predicted user performance extremely closely here, it could safely substitute for expensive user testing. Using modeling and user testing, we learned that a given phone might be advantageous for one type of task but not for another. Each phone had strengths and weaknesses. More importantly, a careful task analysis that outlines the step-by-step procedure to accomplish a task (see Table 6.3) can readily reveal when one device might be superior to another. Changing the Motorola screen size and menu structure to mimic that of the Nokia would have led to a different result in the texting task where the Motorola's better key layout could have advantaged that phone.

   Another interesting finding not discussed in that article was error rate as a function of timeout interval for keypresses to reach target letters. The Nokia had a timeout of 1000 ms, so if you waited more than a second to press the key again, then you would register two letters rather than one. The Motorola used a 1500 ms timeout. The GOMS-predicted margin for error for four repetitions on a key was 241 ms for the Nokia and 508 ms for the Motorola. When we examined errors on entering the letter "s," which required four repeated keystrokes (key contained P, Q, R, S), the Nokia registered about 1 error per trial and the Motorola about 0.4 errors per trial for older adults. For faster-responding younger adults, there were barely any differences in error rates across phones (0.4 vs. 0.5 errors per trial). Again, designing for young adults might selectively disadvantage older adults given the latter's slower rate of responding. GOMS modeling makes these potential design flaws obvious without ever conducting user testing.

6.5   Case Study 3: Using GOMS to Predict Time for Driver Decision Making

Our third example stems from some research conducted for the Florida Department of Transportation at Florida State University: http://safemobi lityfl.com/pdfs/FDOT-BDV30-977-04-rpt.pdf. At the time (circa 2014), automatic red-signal camera systems were being mounted on Florida traffic signals to catch people who tried to speed through intersections after the signal had turned red. That traffic rule violation can lead to very serious "T-bone" crashes where the red-signal-running car crashes into the side of the vehicle with the right of way, compared to less serious rear-end collisions if a lead driver suddenly stops at a changing signal.

Some motorists were convinced that municipalities had shortened the yellow signal duration to make it more likely that a driver would be caught when the signal changed from yellow to red, generating revenue from unfair signal timing settings. Further, generally slower decision time in older motorists as they approached an intersection (see Table 6.2) might systematically disadvantage them with short inter-signal intervals.

The problem for the driver who is approaching an intersection and sees the signal change from green to yellow is whether to continue through the intersection or to halt before it. The driver can be caught in what has been termed the "dilemma zone," which occurs when the laws of physics dictate that a driver can neither make a legal stop before the stop line at the intersection (by braking) nor clear the intersection (by accelerating) before the signal turns red. The driver's decision involves predicting whether it is safe to continue through the intersection based on factors such as vehicle speed, distance from the intersection, intersection width, presence of vehicles or pedestrians in the intersection, and yellow signal duration. Slower responding older drivers have been shown to have a greater dilemma zone: a time of arrival of 3.2–1.5 s away from the intersection compared to younger adults (3.9–1.85 s from the intersection).

6.5.1   Research Questions

Can we predict the dilemma zone for older drivers? Our research team created GOMS models for that decision process, simulating older and younger drivers. Unlike the simpler cases discussed above, with well-understood task decompositions ("keystroke level model"), the dilemma zone case involved significant guess work about the mental processes carried out by a driver confronted with a changing signal. We defined several idealized models: "Cautious Go," "Cautious Stop," and "Semi-cautious Go." The models were idealized because dilemma zones are probably not encountered with very high frequency by drivers (hence not categorized as routine cognitive activities), and drivers are unlikely to be wholly attentive while driving. We drew up the models in advance of testing them to ensure that we provided a fair test. The "Cautious Stop" model is shown in Table 6.4, using parameters shown in Table 6.2 for young and old cycle times.

That model suggested an age difference in decision time of about 770 ms favoring young adults. We then proceeded to test the predictions using a driving simulator to trigger a dilemma zone decision in the scenario by changing the signal from green to yellow as the driver approached the intersection. We looked at driver response times to yellow signal changes for braking responses and adjusted the reaction times to take into account the slower mean driving speed of the older drivers (34 mph vs. 37 mph). Mean adjusted time to press the brake was 2297 ms for younger drivers and 3099 ms for older drivers, for a mean age difference of 803 ms – about 33 ms longer than predicted by the Cautious Stop model. However, the absolute time for the decision was off by about 1000 ms. Real-world driving is unlikely to be as efficient as that in an idealized situation. Further, driving in a simulator might not equate to driving in a real vehicle because people are probably driving cautiously knowing that they are being observed.

6.5.2   How Diversity Influences Performance

Yet again, we see that there were quite significant differences between younger and older drivers, statistically as well as practically, in the

Table 6.4 GOMS modeling for dilemma zone decision making using "Cautious Stop"

Initial state	Operator	Young (ms)	Old (ms)
Signal change in the periphery	Perceptual	100	178
Fixate signal/saccade	Eye fixation	230	267
Decode signal meaning	Cognitive	70	118
Fixate stop line	Eye fixation	230	267
Gauge time to arrival at the intersection	Cognitive	70	118
Determine action	Cognitive	70	118
Dilemma state: cautious driver stopping at yellow			
Saccade to the rearview mirror	Eye fixation	100	178
Determine the traffic situation	Cognitive	70	118
Decision made to stop	Cognitive	70	118
Move foot to the brake pedal	Motor + Fitts	264	485
Press the brake pedal	Motor	70	146
Predicted time:		**1344**	**2111**

estimated time to make an important driving decision. Designing a yellow signal duration time based on younger adult response times would be expected to seriously disadvantage older adults. They would find themselves with almost a second less time to make their decision about whether to stop at the yellow signal or proceed through the intersection. Fortunately, older drivers tend to drive more slowly than younger drivers, probably to compensate for generally slower information processing time (see Table 6.2). Nonetheless, designing signal intervals to suit older driver decision times is likely to be helpful for younger drivers in terms of reducing their dilemma zone.

6.5.3   Design Value

Perception–response time is used in the Manual of Uniform Traffic Control Devices (MUTCD) to design many aspects of the road environment: from where to place warning signs to signal durations. The assumed perception–response time for yellow signals is 1 s, meaning that 1 s is considered adequate time for a driver to perceive a signal change and respond appropriately. Older drivers are likely not well served by this assumption, at least insofar as setting yellow signal durations where they took about 3 s to press the brake on deciding to stop. Even younger drivers took about 2.3 s. Timing of traffic signals takes both safety and traffic flow factors into account, and hence represents the usual case of compromises driving the design process.

Unfortunately in recent years (2015–2017), older driver fatal crash rates have been increasing (e.g., Insurance Institute for Highway Safety data). As well, general commuting times have shown increases: 26.4 minutes in 2017 compared to 25.4 minutes in 2012 (American Community Survey data). However, traffic engineers might want to readjust weighting to emphasize safety over traffic flow (Hauer, 2019), given the increasing numbers of older drivers on our roadways.

6.6   Conclusions

In summary, we have explored the use of modeling techniques, such as Fitts' Law for predicting movement time, and GOMS modeling for estimating task completion times on routine tasks. We discussed three cases. The first explored the relative advantages and disadvantages for older and younger users of interface elements as a function of their spacing for a touchscreen device. The second explored the relative advantages and disadvantages of feature phones for different tasks, dialing, and texting, for both younger and older adults. The final case discussed modeling dilemma zone decision times for older and younger drivers when they approach intersections with a yellow signal. In each case, there were clear benefits shown for modeling performance.

Ideally, the design process can make use of both types of modeling, where appropriate, and user testing to validate the modeling. However, when user testing might be too costly to conduct (time, money), modeling can be good enough to guide decisions such as choosing between two different designs.

References

Card, S. K., Moran, T. P., & Newell, A. (1983). *The psychology of human-computer interaction*. Lawrence Erlbaum Associates.

Fitts, P. M. (1954). The information capacity of the human motor system in controlling the amplitude of movement. *Journal of Experimental Psychology, 47*(6), 381–391. doi:10.1037/h0055392.

Hauer, E. (2019). Engineering judgment and road safety. *Accident Analysis & Prevention, 129*, 180–189. doi:10.1016/j.aap.2019.04.022.

Jastrzembski, T. S., & Charness, N. (2007). The Model Human Processor and the older adult: Parameter estimation and validation within a mobile phone task. *Journal of Experimental Psychology: Applied, 13*(4), 224–248. doi:10.1037/1076-898X.13.4.224.

Morrow, D. G., & Rogers, W. A. (2008). Environmental support: An integrative framework. *Human Factors, 50*(4), 589–613. doi:10.1518/001872008x312251.

Rogers, W. A., Fisk, A. D., McLaughlin, A. C., & Pak, R. (2005). Touch a screen or turn a knob: Choosing the best device for the job. *Human Factors, 47*(2), 271–288. doi:10.1518/0018720054679452.

Verhaeghen, P. (2014). *The elements of cognitive aging: Meta-analyses of age-related differences in processing speed and their consequences*. Oxford University Press.

Welford, A. T. (1977). Motor performance. In J. E. Birren & K. W. Schaie (Eds.), *Handbook of the psychology of aging* (pp. 450–496). Van Nostrand Reinhold.

Additional Recommended Readings

A detailed discussion of Fitts' Law can be found online at https://www.yorku.ca/mack/GI92.html.

A Fitts' Law calculator can be found online at https://codepen.io/gsus/details/QyKBGy.

A keystroke level modeling (KLM) approach by Jeff Sauro is outlined at: https://measuringu.com/predicted-times/.

Some spreadsheets for GOMS modeling and Fitts' Law are on the CREATE web site: https://create-center.ahs.illinois.edu/instruments.

chapter seven

Designing Instructional Support

7.1 Introduction

Anybody know of any completely "intuitive" technology products? You can probably find them in the same department that houses (mythical) unicorns. Once designers realize the range of potential users for a product and that there may be novice users in that group, the need for instructional support becomes obvious, even for second- and third-generational variants of existing products. How can we design instructional support for aging users? As Chapter 6 GOMS parameters illustrated, older users undergo normative changes in information processing capabilities supporting perception, cognition, and psychomotor performance. Slowing in information processing rate, with older adults taking 1.5 to 2 times longer than younger adults for many cognitive operations, provides an important constraint for learnability and memorability, two important facets of product usability (see Chapter 4). Age-related diminished capabilities in senses such as vision and hearing that support perceptual processing provide important constraints on how to design multi-media supports such as manuals and videos. In this chapter we focus on three case studies from CREATE research on best practices for designing instructional support.

7.1.1 Support as an Underutilized Part of the Design Process

It seems to be the case that design of instructional support is generally a low priority, at least judging by the materials often supplied with lower-cost technology products. Even high-end products such as smartphones typically only have a brief start guide packaged with the phone, though there are usually links to an online manual. Is that instructional support adequate? Figure 7.1 shows a population-representative survey done by the Pew Internet and American Life group in 2015, asking people if they needed help with using a new electronic device. They were given the statement: "When I get a new electronic device, I usually need someone else to set it up and show me how to use it." They were asked how well that statement describes them, with the top two categories being *somewhat well* or *very well*.

So we can tentatively conclude that instructional support is generally inadequate, with nearly three-quarters of those aged 65+ requiring additional personal help. Why might that be the case?

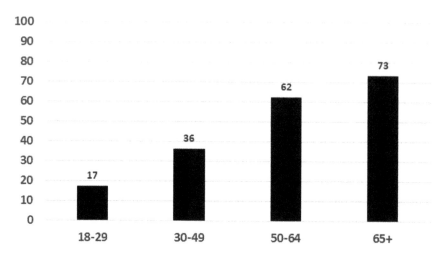

Figure 7.1 Percentage of Americans self-describing that they need help from someone with setting up and using a new electronic device by age group: bars represent the sum of the percentage of people who felt that this statement describes them well or very well. Data from http://pewinternet.org/2017/05/17/tech-adoption-climbs-among-older-adults/

7.1.2   Older Adults Do Read Manuals

A common phrase that tech support has been known to proffer to product users who found themselves in trouble was "RTFM," as in Read The Frigging Manual. That was rude but sensible when products came with detailed manuals explaining their functionality. However, in the very competitive consumer tech industry that tends to show shrinking product margins over time, companies progressed from detailed printed manuals to manuals on floppy disks and then on CDs, flash drives, and now web-based/online manuals and tutorials (e.g., YouTube videos).

Online manuals and tutorials make economic sense as they greatly reduce product cost. However, older adults are less likely to be on the internet. As of this writing 25% of Americans aged 65+ are still not using the internet, as assessed by a question such as, "do you use the internet or email at least occasionally," and are thereby less likely to be able to access internet resources easily. Further, we have found that older adults often prefer to read paper manuals to find out how to use a product. As stated above, they often need to get help from a human, but toll-free telephone support lines often lead to their interaction with a non-native English speaker whose accent may be difficult to comprehend (e.g., support service centers based in Mexico, India, and Pakistan). If you want to market your product to aging adults, at this point in history, providing a paper manual may be a selling point, along with the provision of locally based telephone support options listed in the accompanying materials.

7.1.3   Iterative Design for Instructional Materials

We have been arguing for a user-centered design approach for products and services. Instructional materials are no exception to the rule. Ideally: prepare your materials, test them with representative (older) users, and redesign them. Then test the revised materials with your intended users to see if your changes have improved their performance. If not, repeat the cycle until performance based on the product and the instructional materials leads to adequate performance. Of course, there are potential interactions possible, because user testing may determine that changes are needed in product design, instructional materials, or both. We now turn to some case studies that illustrate different approaches to designing instructional materials.

We start with telephone menu systems as they represent an example of the importance of environmental support within instructional design. Because of the heavy working memory load involved in tracking aurally provided alternatives within a large menu structure, finding effective ways to present information to the user, and to allow them to backtrack, is critical.

7.2   Case Study 1: Design of Instructional Support for Telephone Menu Systems

7.2.1   Research Questions

At the time of the original study (circa 2001), internet-based services were not very common and wired telephone systems were the main way for companies to provide product support. Even today, a common task for most older adults is refilling prescriptions, which they often do via pharmacy-provided voice-based telephone menu systems. Telephone voice menu systems are good examples of a joint product and instruction set because the hierarchy of options for the user is in a sense the product, and the specific wording of the menu entries is the help system. Assuming that you have to navigate through the tree of options to find the relevant node to transact your business, instead of immediately pressing 0 to bypass all the options to be connected to a live operator, you are faced with a task that can be very frustrating or relatively simple. The research question under investigation concerned how best to support navigation in voice based telephone menus for use by older adults.

We investigated (Sharit, Czaja, Nair, & Lee, 2003) several factors in the successful use of a voice-based telephone system in a study of young, middle-aged, and older adults. The first experiment looked at the speech rate for the recorded voice prompts. However, we focus here on the second study that assessed two types of instructional support: a screen phone or a graphical aid for the simulation of two menu systems (banking, an electric utility). The main concern with such voice-based menu structures is that they place a heavy load on working memory, which shows significant age-related declines. A typical menu will present multiple choices at the entry point of the hierarchy/inverted tree, and the person must remember the choices to be able to select the right path toward the goal state (e.g., the bottom options in Figure 7.2). Each further press of a button will take you to another set of voiced options, and if you discover that you are on the wrong path, you need to find a way to climb the tree back to an earlier node. It is a near-perfect example of searching through a problem space to reach your goal (Newell & Simon, 1972).

The first experiment that tested for speech rate effects found none, though a significantly greater number of errors had been committed by older adults (compared to middle-aged and younger adults). The authors determined that these errors were associated with older adults being unable to access the final submenu (bottom of the tree), or if they did find it, errors were committed either in entering completely correct information or in omitting needed information. They also found that working memory and attention measures were strong predictors of performance.

Figure 7.2 Screen phone from Sharit et al. (2003).

One problem with wired telephone handsets is that because many of them have the buttons on the handset, people needed to switch between listening, then turning over the phone to find and press the relevant button, and then putting it back up to their ear. The same problem arises with mobile phones in the case of not using a speaker-phone function or earbuds. Task switching is an example of an atten-tion-demanding task because of the need to switch between mental operations such as comprehending and remembering where you are in the menu structure, then doing a visual search and pressing a but-ton, and then returning the handset back up to the ear to listen to the next option. To counteract the working memory and attention demands, Sharit and colleagues (2003) tried to provide "environmental support" (Morrow & Rogers, 2008) to the user by providing them either with a screen phone (see Figure 7.2) that also separated out the speaker from the buttons or with a graphical aid, a diagram showing the layout of the menu structure (as seen in Figure 7.3 for the banking task). The screen phone can reduce memory load by visually displaying the menu choices simultaneously for a given level of the search tree. That is, it provides redundant information to the user about where they are in the menu structure. It also eliminates the interference that might arise when the person must search for and press a button after turning the hand receiver over to access the keypad.

In this study, all three conditions tested involved using the screen phone: screen phone, graphic aid (two environmental support conditions), and a control condition. Only the screen phone condition activated the screen function that showed a menu level, and in the control condition there was no screen information. People in the environmental support conditions were given three practice problems to familiarize themselves with the support tools and then completed a questionnaire after each call to rate the extent to which they used the support tool. Three age groups

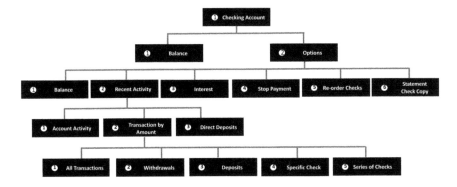

Figure 7.3 Graphical aid showing the hierarchical menu flow structure.

(young, middle, old), with 38 participants in each, completed the same task as in the first experiment.

The pattern of results was complex, given the different outcome measures that included overall performance, time per call, repeated requests, and navigational efficiency. Also, the analyses split problems into low and high complexity and analyzed separately for bank and electric utility problems as well as for age effects. Mean performance scores improved most with the screen phone across all age groups compared with the control condition. The graphical aid seemed to benefit older adults more for the electric utility problems. Analyses of the self-rated use of the support device showed that screen phone use exceeded graphical aid use.

7.2.2   How Diversity Influences Performance

Age is an important individual difference factor in performance with voice menu systems. However, age appears to exert its influence through age-related changes in working memory and attention. Although all age groups performed about the same with different speech rates (Experiment 1), older adults performed somewhat worse than or about the same as middle-aged adults, with both age groups performing somewhat worse than younger adults. Importantly, having environmental support in the form of screen information or a graphical aid displaying the menu structure had slightly different impacts on different age groups. What works best for one age group does not necessarily work best for another.

7.2.3   Design Value

It seems unlikely that companies will provide users with screen phones or graphics of menu structures for their voice-based telephone menu systems, even if training with such supports would improve older adult performance. Although voice-based menu systems are being supplemented (and replaced) by internet-based visually guided menus, design decisions about menu structure remain. Further, many users, including some older ones, are choosing to access internet services via smartphones. So attention to the fit between user capabilities and environmental/product demands will continue to be necessary. Providing environmental support for user activities is good business practice.

As an example, smartphones provide the user with many of the advantages of screen phones in terms of providing text along with both visual and auditory guidance. Plus, accessibility features in these phones provide the opportunity to increase text size beyond that shown on an old-fashioned screen phone and to increase sound volume (or pipe it directly into the ear with earbuds or headphones). Providing product support through smartphone apps that take advantage of such features can give users the best features of both visual and aural menu systems.

The finding that reducing working memory demands through environmental supports points to the potential utility of web-based and app-based menu systems. As an example, people in the United States typically refill some prescriptions either by phoning and using a voice-based menu system or by going to the pharmacy's web site and ordering from there. Web-based, consumer-facing systems can present navigation challenges, particularly for older users (Czaja, Sharit, & Nair, 2008). Nonetheless, one advantage is that at any one level in the menu hierarchy, working memory load is reduced, similar to the screen phone situation, by having the alternatives visible for scanning rather than having to hold the alternatives in memory for a sequentially provided auditory menu system. Giving older adults redundant choices (voice, web) may be the optimal solution for instructional support in this case. For either option, user testing can reveal ways to improve performance.

7.3 Case Study 2: Blood Glucose Meter Usability

7.3.1 Research Questions

Imagine that you have visited your physician complaining of symptoms such as urinating frequently, feeling very thirsty and hungry, losing weight despite eating more, suffering from fatigue, and having blurry vision. The physician runs some tests and diagnoses you with type 2 diabetes. You are told that you need to check your blood glucose levels daily and purchase a blood glucose meter to do this. The meter you selected suggests that carrying out this procedure is "as easy as 1-2-3." However, it turns out to involve a bit more complexity, about 52 steps to operate (Rogers, Mykityshyn, Campbell, & Fisk, 2001). This task analysis case study is detailed in Chapter 4. Given this complexity, how might the designer best instruct you on its operation?

We tried to answer this question by designing new training materials, specifically a new manual and a new video, to help the user (Mykityshyn, Fisk, & Rogers, 2002). The materials relied on several training principles, such as writing instructions that encourage integration processes by the user, hence leading to better performance, also making use of visual training that is already well integrated to begin with, hence reduces working memory demands. We were looking at the comparative effectiveness of these two approaches to instructional design and compared 30 older and younger adults who had never used a blood glucose meter (and who did not live with someone who used the device). The task was to calibrate the meter, a critical step in using the meter properly to test blood glucose levels.

We first carried out a detailed task analysis of the steps needed to carry out the procedure. That process ensured that readability levels were appropriate, that numbered steps were given along with an advanced organizer to help conceptualize the procedure, and that redundant information was provided (text with figures, spoken words with visual images). Critical information was repeated. The main difference between conditions was the format of presentation: text with pictures or videos. The user manual comprised eight legal-size pages based on manufacturer materials and experienced user input. The manual included a schematic drawing of the blood glucose meter and its main components, as well as step-by-step instructions on how to perform the three tasks required to calibrate the meter. Pictures accompanied the text instructions for each step.

A transcript from the manufacturer formed the basis for the video. It was filmed from the perspective of the user and showed step-by-step how to accomplish the task with close-up views of the meter. Each calibration sub-task started with an image providing the title of the task such as "performing a glucose control solution test" and a description of when to perform it, along with accompanying text provided in bullet form on the screen. That is, the steps were provided in advance of being demonstrated, to parallel the format of the manual. The video was about eight minutes long and distributed on a videocassette.

Participants in each age group were randomly assigned to the two conditions and instructed to study the materials until they felt they could perform each of the calibration tasks. Materials were removed, and they were then tested on two complete calibrations of the meter. Workload and confidence ratings were obtained after each calibration, and a knowledge test was given after the second calibration. The participants were tested again two weeks later without additional training. In the first session, study time was initially near equivalent between younger and older groups on the manual (14 minutes), but older adults spent more time than younger ones with the video (18 minutes versus 12 minutes – they were allowed to replay it as needed).

Calibration accuracy immediately after instruction revealed that younger adults were more accurate than older ones, the video yielded better performance than the manual, and the older adults benefited differentially from the video. Older adults' performance in the video condition was equivalent to younger adults in either video or manual conditions. The time taken to do the calibration showed an age effect, with older adults being slower than younger ones. The condition did not affect time. However, there was an age-by-calibration interaction, showing that older adults improved more than younger ones on the second calibration task. After two weeks, older adults showed a steeper decline in accuracy than younger ones, though not differentially across training conditions. A similar effect was seen for calibration time.

The knowledge test taken after training showed only an age effect, with older adults' performance lower than younger adults. However, knowledge tested two weeks later showed a differential benefit for video training in the older group, though there was also the main effect of the age group. Similarly, for workload ratings, older adults rated workload differentially higher in the manual condition compared to the video condition, though both age groups experienced lower workload with the video training.

7.3.2 Design Value

A crucial finding was that both younger and older adults were significantly below 100% accuracy. For a medical device that provides critical information about blood glucose levels, less than perfect performance is a worrisome finding. However, the design process here was the first step of an iterative cycle that would normally include redesign and retesting to ensure that 100% accuracy was attainable. Also, in real-world settings, users would retain access to the instructional materials as they performed the calibration tasks, so we do not know if the accuracy would have reached performance ceiling, had they been able to refer to the materials, whether in the form of a manual or a video. Nonetheless an important finding was that the video differentially helped older adults' calibration accuracy. In fact, the well-designed video virtually erased performance differences between the age groups. Today's typical YouTube instructional video could benefit from a similar careful approach to instructional design.

In conclusion, it appears that a video format can differentially benefit older adults relative to a printed manual. However, it is critical to test instructional materials under realistic testing conditions and perhaps to warn users about overconfidence in the sense that allocated study time in this study was not commensurate with an errorless performance for either younger or older adults. Generally, though, older adults tend to be less confident than younger ones, which is a good bias to have when it comes to using medical equipment.

7.4   Case Study 3: Florida DOT Tip Card Project

7.4.1   Research Questions

Mobility is critical to independence for aging adults. However, the road environment is always changing and road users – drivers, cyclists, and pedestrians – need to be informed about how to navigate safely as changes are introduced. For instance, in some Florida cities, new signals were introduced to facilitate traffic and safe pedestrian flow such as the flashing yellow arrow at four-way street intersections and rectangular rapid flashing beacons at pedestrian crossings. Further, the rules for driving safely through roundabouts and for making a safe right turn on red are not necessarily intuitive. How can educational tip cards be optimized to inform older drivers and pedestrians? What are good guidelines for designing printed materials to inform road users about roadway changes? We (Charness et al., 2017) generated guidelines for brief instructional materials such as two-sided tip cards and multi-panel brochures and provided some templates for two-sided cards.

What are the appropriate criteria for evaluating whether one tip card design is superior to another? Typically, human factors user testing examines facets such as learnability, efficiency, memorability, errors, and satisfaction. We focused primarily on the first four facets. In a series of experiments, we compared existing tip card designs to enhanced versions that followed guidelines generated following a literature review. The review aimed at understanding how best to design materials for older adults that focused on age-related changes in attention, encoding, comprehension, attitudes, and motivation. Generally, when people read such public service materials they look at them fairly briefly (a few minutes), so it is necessary to grab their attention, facilitate encoding of critical information, support comprehension processes, and then persuade them to adhere to the instructions by inducing positive attitudes toward the message and motivating them to go along with it. Figure 7.4 provides the checklist generated to guide the production of such materials.

Following the checklist creation, enhanced tip cards were generated and then tested against the existing tip cards, first for learnability by assessing comprehension with free recall and then multiple-choice questions immediately following reading (front and back were displayed sequentially on a screen) in a total of 307 younger, middle-aged, and older adults. Two signal types were instructed: flashing yellow arrow (FYA) and rectangular rapid flashing beacon (RRFB). The original card was contrasted with an enhanced card in two appeal types (positive, negative) for each signal type (FYA, RRFB).

For learnability, there were no advantages for the enhanced cards or for the type of appeal. However, for efficiency, assessed as reading times on cards, there was a large advantage for the enhanced cards (31 s compared to 46 s reading time for the control cards). That is, people processed enhanced cards more quickly while showing equivalent memory for the contents.

The next study with 158 middle-aged and 161 older adults investigated memorability by having people read the cards and be tested immediately or come back for testing a week later with memory tests as well as a performance test that asked people to look at a scene shown either as a photo of a real intersection or a "Google Sketchup" image (3D scene) and

Checklist for Design of Tip Cards and Brochures for Aging Road Users		
Factor	**Advice**	**Check**
Legibility		
	Font size minimum of 12-14-point (x-height)	
	Serif font if large, otherwise sans-serif	
	Prefer bolded text, particularly for headers	
	Avoid decorative font	
	Mixed case for body text except where emphasis is needed then uppercase	
	High enough contrast that can be read at <40 cd/m²	
	Prefer black on white or white on black text	
	Consider colored text or backgrounds for emphasis but avoid blue/violet	
	Left-justify text for passages	
	Double-space text when possible	
	Limit line lengths to 50-65 characters for brochures	
	Avoid wrapping text around pictures and illustrations	
	Avoid glossy material for cards and brochures	
Pictorial Materials		
	Add pictures to text to convey complex instructions	
	Prefer high resolution photos to convey real-life events	
	Prefer high quality illustrations when conveying detailed information	
	Caption pictorial materials that are not easy to interpret	
	Try to use culturally relevant illustrations	
Layout		
	Provide key information first (top)	
	Use bulleted lists to break up paragraphs of text	
	Use color to make the material attractive and engaging	
	Use headings and subheadings to create visible sections	
	Try to keep 10-35% of the page as white space to reduce clutter	
Comprehension & Memory		
	Try to cover only one general topic per card	
	Chunk information and use short sentences	
	Present 6 or fewer chunks of information in a section	
	Use active voice and avoid passive and negative phrases	
	Avoid jargon by using everyday language	
	Aim for a Flesch-Kincaid score of Grade 8 education or lower	
	Visuals should support imagining the actual road environment	
	Keep alternating phase representations close together to support integration	
	Focus on actions for road users to take	
	Encourage simulation of the target behaviors	
	Encourage self-testing of memory for the target behaviors	
	Encourage self-reference by using terms such as I rather than driver	
Attitudes		
	Consider an emotional appeal to facilitate attention, memory, and positive attitudes toward the behavior	
	Try to enhance self-efficacy of the road user by using positive appeals	
	Remind road users in a non-threatening way about regulations	
	Consider generating alternate forms of the material to maintain attention	
	Create an electronic version for distribution through social media	
	Consider reinforcing information with road signs, ads, press releases	
Motivation		
	Try to specify implementation intentions rather than goal intentions by suggesting concrete steps to adhere to the regulation	
	Use consistent layout and logos to brand materials to enhance credibility	

Figure 7.4 Checklist for designing tip card materials.

to make a rapid decision about whether the driver should stop, yield, or go. The scene contained either an FYA or RRFB signal. Conditions included no tip card, an enhanced card, or the control card. The participants could study the card for about four minutes. Memorability this time was unaffected by the type of card, age, or delay. However, performance on the speeded decision task was slower for the older adults compared to the middle-aged ones in the immediate testing condition. After a delay of one week, however, there was no age effect on response speed. There were no age effects, condition effects, or delay effects on the accuracy of response.

In the next study, aimed at assessing errors in roadway performance, 40 middle-aged and 61 older drivers were exposed to an enhanced relevant tip card for up to two minutes: FYA or turning right on red (ROR), and an irrelevant tip card: RRFB and roundabouts. Reading times were unrelated to the type of card or age. Drivers were then asked to drive in a simulator after a short acclimation task. The scenes they encountered contained an FYA with an oncoming vehicle (meaning that driver should yield, not turn, until the oncoming car cleared the intersection), and then a ROR situation with a permissible turn after an oncoming vehicle cleared the intersection. For FYA, neither having the correct tip card, nor age, nor the presence of an oncoming vehicle affected wait time for making the turn. For ROR, having the correct tip card shortened the wait time for permissible turns, and when a no-turn-on-red sign was present, older drivers waited longer than middle-aged ones. (Drivers were quite conservative in these scenarios.) Nonetheless, a well-designed tip card aided decision making for making right turns at a red signal.

7.4.2 Design Value

By adhering to human factors guidelines (see Figure 7.4) when designing a tip card, a very large advantage was seen for adults of all ages on the time to read/comprehend the enhanced tip card information relative to the standard card. Good design principles led to faster reading at no cost to comprehension accuracy and benefited all age groups.

7.5 Conclusions

Instructional support is too often an afterthought in the design process for products and services. Older adults report needing more instructional support than younger age groups, but as Figure 7.1 illustrates, everyone can benefit from better instructional support. The case studies demonstrated some useful approaches to user testing of instructional support materials, emphasizing learnability, memorability, efficiency, and errors, though user satisfaction should not be overlooked. Assessing errors and error types is of particular importance for the use of medical devices.

The cases indicated that different formats had advantages and disadvantages. For instance, providing visual environmental support helped older adults navigate voice-based menu systems more successfully, an example of the value of redundant information channels. Video provided superior support to a static manual, in part because a video is inherently better for portraying action steps than a sequential set of static pictures. For printed tip cards, attention to legibility and comprehensibility guidelines promoted much faster reading than for the standard materials, while maintaining comprehension levels. Generally, the same guidelines for designing products hold for designing instructional materials.

References

Charness, N., Boot, W., Kaschak, M., Arpan, L., Cortese, J., Clayton, R., Stothart, C., Roque, N., Paedae, B., & Barajas, K. (2017). *BDV30 977-15 Human factors guidelines to develop educational tip cards for aging road users.* Retrieved from http://safemobilityfl.com/pdfs/FDOT-BDV30-977-15-rpt.pdf.

Czaja, S. J., Sharit, J., & Nair, S. N. (2008). Usability of the Medicare health web site. *JAMA, 300*(7), 790–792. doi:10.1001/jama.300.7.790-b.

Morrow, D. G., & Rogers, W. A. (2008). Environmental support: An integrative framework. *Human Factors, 50*(4), 589–613. doi:10.1518/001872008x312251.

Mykityshyn, A., L., Fisk, A. D., & Rogers, W. A. (2002). Learning to use a home medical device: Mediating age-related differences with training. *Human Factors, 44*(3), 354–364. doi:10.1518/0018720024497727.

Newell, A., & Simon, H. A. (1972). *Human problem solving.* Prentice-Hall.

Rogers, W. A., Mykityshyn, A. L., Campbell, R. H., & Fisk, A. D. (2001). Only three easy steps? User-centered analysis of a "simple" medical device. *Ergonomics in Design, 9*(1), 6–14. doi:10.1177/106480460100900103.

Sharit, J., Czaja, S. J., Nair, S., & Lee, C. C. (2003). Effects of age, speech rate, and environmental support in using telephone voice menu systems. *Human Factors, 45*(2), 234–251. doi:10.1518/hfes.45.2.234.27245.

Additional Recommended Readings

Czaja, S. J., & Sharit, J. (2013). *Designing training and instructional programs for older adults.* CRC Press.

Tsai, W. -C., Rogers, W. A., & Lee, C. -F. (2012). Older adults' motivations, patterns, and improvised strategies of using product manuals. *International Journal of Design, 6*(2), 55–65.

chapter eight

The Personal Reminder Information and Social Management System (PRISM)

As discussed in Chapter 1, the focus of this book is on design for aging adults, a critically important and growing user group. The central goals of the book are to provide illustrative case studies of design problems across of variety of domains that have been undertaken by the Center for Research and Education on Aging and Technology Enhancement (CREATE) and describe the tools and techniques used to address these challenges. In this chapter, we present an example of a software system developed for older adults by the CREATE team, the Personal Reminder Information and Social Management System (PRISM). PRISM was intended to support social connectivity, prospective memory, knowledge about topics and resources, and entertainment. The PRISM system encompassed a software program, a training package, and user support tools (user manual and help card). The development of PRISM involved the use of most of the human factors usability tools and methods discussed in Chapter 4 and highlighted in the case studies described in other chapters. Our intent in this chapter is to present a case study that begins with the inception of a system to the evaluation of the system in a multi-site field trial to illustrate how an iterative, user-centered design approach can be used to design a system that accounts for the needs, abilities, and preferences of a diverse older adult user group.

The chapter begins with an overview of the research issue, followed by a description of the PRISM system and the process used to develop and evaluate PRISM. We also highlight some of the challenges we encountered to demonstrate that design is a complex process that requires adaptability and compromise. In the design of the PRISM system, we broadly followed the design thinking process outlined in Chapter 3 (see Figure 3.1), which involves three main activities – understand, explore, and materialize – each of which has associated steps. In general, design thinking is a non-linear, iterative process, which involves understanding users and user needs, defining user problems, creating innovative solutions to these problems, and building prototypes and testing these solutions (see Figure 8.1).

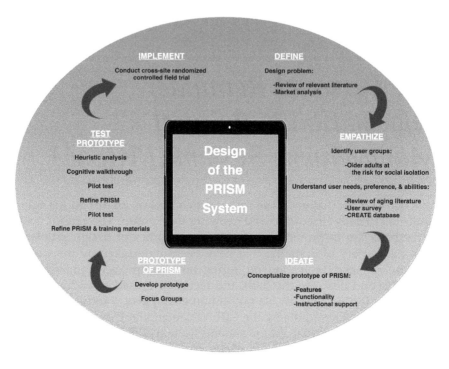

Figure 8.1 The PRISM design process.

8.1 The Issue: Social Isolation, Loneliness, and Older Adults

Social isolation and loneliness are worldwide problems that affect many older adults. Recent data suggest that about 24% of adults aged 65 or older are socially isolated (Cudjoe et al., 2018), and data from the NHATS survey of a population-based sample of older adults in the United States aged 62–91 indicated that 19% of the respondents suffered from loneliness and an additional 19% felt lonely often or some of the time (Hawkley, Kozloski, & Wong, 2017). Problems with isolation and loneliness are especially prevalent in adults in the older cohorts; those who live alone or in rural locations; or who have low income, low education, or a disability.

"Social isolation" refers to a physical lack of separation from others and the lack of social support, whereas "loneliness" reflects a subjective state of being dissatisfied with social relationships. Reasons for social isolation and loneliness are varied and include geographic separation of families, changes in life roles such as retirement, diminished financial resources, loss of partners and friends due to death or disability, illness, functional declines, and cognitive declines. Irrespective of the reasons, social isolation and loneliness represent important public health issues and are associated with poor quality of life, reductions in physical and mental health, and declines in cognition. The prevalence of isolation and loneliness among older adults, combined with the potential impact on an individual's health, has led to a focus on developing strategies to enhance social connectivity and engagement to prevent loneliness among older people.

In today's digitized world, investigators have begun to investigate how technology applications can be used to foster opportunities for social engagement among older people. Some studies have found that providing older adults with access to technology applications has no impact on well-being (e.g., Dickinson & Gregor, 2006; Slegers, van Boxtel, & Jolles, 2008). However, others have shown that such access improves quality of life and reduces feelings of loneliness (Choi, Kong, & Jung, 2012; Cotten, Anderson, & McCullough, 2013). For example, Cotten and colleagues (Cotten et al., 2014) analyzed data from the Health and Retirement Survey, a longitudinal household survey for the study of retirement and health in the United States, and found a positive association between internet use to mental health and that use of the internet may decrease loneliness and depression. The mixed findings in the literature may be due to a variety of factors including variation in study populations, technologies examined, and outcome measures. Also, only a paucity of studies are systematically rigorous (e.g., include a control group); included large, diverse samples; or used systems that were specifically designed for older adults.

Given our interest in problems of social isolation among older adults and in examining the potential of Information and Communication Technologies (ICTs) for addressing issues confronting older adults, we designed and evaluated a software system using an iterative user-centered design approach, the PRISM system. As noted, PRISM was intended to support social connectivity, prospective memory, knowledge about topics and resources, and entertainment. In the following section we present a description of the PRISM system.

8.2   Description of the PRISM System

The PRISM system included internet access (with a menu of vetted links to sites for older people such as http://go4life.nia.nih.gov), an annotated resource guide, a classroom feature, a calendar feature, a photo feature, email, games (see Figure 8.2), and an online help feature. Features were accessed with a single click on the feature name on the sidebar menu, and the main categories of features were arranged using tabs that appear at the top of the screen (see Figure 8.3).

The classroom feature was dynamic and contained scripted information, vetted videos, and vetted links to other sites on a broad array of topics (e.g., cognitive health, traveling tips, nutrition). New material was

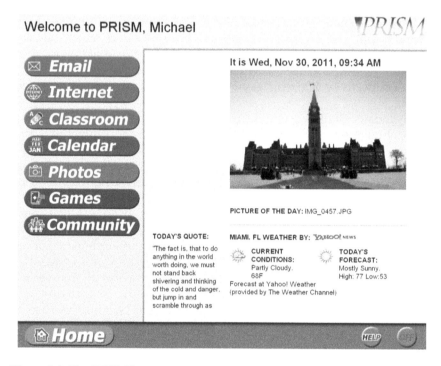

Figure 8.2 The PRISM home screen.

placed in the classroom every month and remained in the "classroom library." The email feature had a "buddy component" intended to foster social connectivity. Upon enrollment in the trial, participants assigned to the PRISM condition were asked if they wished to be a "PRISM buddy." If they agreed, their email address was placed in the "PRISM Buddy" tab of the email feature, as well as a few key words describing their hobbies and interests. The photo feature was preloaded with an album created by the research team, and participants were able to create their own albums and share photos. In addition, an online help feature was available. A software monitoring program was developed to monitor system use that parsed usage on a daily basis. An automatic email message was sent to the study site coordinators if a participant had not used the system for more than seven days, who then contacted the participant to determine the reason for non-use (e.g., technical difficulties). The homepage contained the date and time, the weather, a scenic picture, and a motivational/inspirational quote of the day. The research team pre-selected pictures and quotes. Users accessed PRISM by turning on the computer; there were no login requirements.

As described later in this chapter, we evaluated the PRISM system with a large and diverse sample of older adults at risk for social isolation

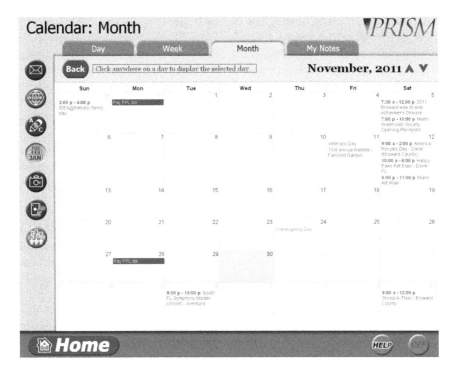

Figure 8.3 An example of a screen from the calendar feature.

and who had minimal or no computer experience from across three cities in the United States in a multi-site randomized controlled field trial (RCT). We chose this user group as we determined that they would serve to benefit from the use of PRISM.

8.3   Design of the PRISM System: Understanding

8.3.1   Defining the Design Problem

Following the design thinking framework in Figure 3.1, we began with *understanding*, which involves *defining* the design problem and *empathizing*, understanding the needs and characteristics of the intended user group. The challenge involved developing a technology system that fostered social engagement among older adults and that was useful, usable, and acceptable to users. We began with a review of the literature that was focused on examining the impact of technology applications on outcomes such as quality of life, social support, and social engagement among older adults. Our intent was to summarize what was known, the populations and technologies studied, factors associated with design and implementation challenges, and factors that influenced intended outcomes (e.g., quality of life). For example, Berkowsky et al. (2013) examined factors that influenced attitudes toward and readiness to use ICTs among older adults in assisted living facilities and highlighted the importance of designing training programs that meet the needs and abilities of the trainee population. We also gathered information about the features of technology systems that were valuable to older adults. The literature indicates, for example, that the association between technology applications and positive outcomes for older adults is related to improvements in their ability to communicate with family and friends and the ability to access news and other useful information (Choi et al., 2012). These findings helped to identify features such as email and internet access that should be included in the PRISM system.

At this stage, we also conducted a market analysis to identify other available systems or technology applications that might be competitive or complementary to our planned system. In doing so we identified the Big Screen Live platform, a software-as-a-service application that was designed to provide easy access to internet services such as email, photo sharing, news, web browsing, games, and simplified online shopping (Carousel Information Management Solutions Inc.). We determined that given our resources and time constraints, it would be more efficient to adapt the Big Screen Live platform to meet our needs as opposed to developing a system from the ground up. Initially, we met with the representatives of the Big Screen Live team to learn about the system to understand the degree to which it could be modified to meet our needs. We developed a work plan with the Big Screen Live team that established timelines, responsibilities, and allocation of activities between Big Screen Live and CREATE. This coordination was an essential aspect of the process as we sought to ensure that everyone had a mutual understanding of the PRISM development process.

8.3.2 Empathize: Understanding Our Intended User Group

Defining the intended user group of a product or system and understanding their needs, preferences, characteristics, and abilities are fundamental components of the design process. Our target population was adults aged 65 and older, who were "at risk" for social isolation. Initially, as we were eventually going to evaluate PRISM in a cross-site field trial, we had to define what we meant by "at risk for social isolation." We decided to concentrate on one aspect of social isolation – living arrangement – and thus our focus was on older adults who lived alone. We recognized, however, that living alone does not necessarily mean that a person is socially isolated. Thus, we expanded our definition of "at risk for social isolation" to include constraints on working or volunteering (<5 hours per week) and engaging in social activities (<10 hours per week at a senior or community center or other organization).

Additionally, we had to consider the characteristics and needs of this user group. As discussed in detail in Chapter 2, older adults are very heterogeneous and vary on a number of dimensions such as culture/ethnicity, gender, socioeconomic status, health, cognitive status, literacy, and functional abilities. In addition, there are normative age-related changes in abilities, for example, declines in vision and manual dexterity, that have implications for system design. We considered these characteristics from two perspectives: accommodating them in the design of the PRISM system and identifying characteristics that would preclude someone from using the PRISM system. For example, in our design of PRISM we focused on the visual dimensions of the display screens such as font size and contrast between labels and the background to accommodate age-related changes in vision. We also attempted to minimize screen clutter and maintain consistency in the functionality of features to accommodate age-related changes in cognition. PRISM involved reading text from a screen, using a keyboard and a computer mouse, learning new information, and understanding oral and written instructions. Thus, we needed to confine our user group to older adults who could see text on a computer screen with or without corrective lens, could hear oral instructions, could understand written and spoken language at a sixth-grade reading level, were non-cognitively impaired, and were able to use a keyboard and mouse.

We gained further insight into the needs of our user group through a survey study conducted at two of our sites (Atlanta, GA, and Tallahassee, FL). The sample included 321 participants (57% were female) who ranged in age from 60 to 93 ($M = 74.62$; $SD = 5.98$). Most of the survey respondents (88%) were active users of computers. We chose to focus on people with computer experience, as the survey was intended to gather information about the importance of various activities (e.g., socializing) to quality of life, the value of having access to computers and the internet, and features and information topics that would be of potential value to older adults. We learned that email, sharing photos, and the ability to search the internet were features of high importance and that access to information related to resources (e.g., Social Security), health, finances, computer skills, and vacation activities was also rated as important among our respondents. The implications for PRISM were that the system should provide access to "useful" information (e.g., health and wellness, news sources), support multimedia communication (e.g., email, photo sharing), and support education and learning and that the training should focus on the benefits of using PRISM as well as on the procedures of how to use PRISM.

8.4   Design of the PRISM System: Exploration

8.4.1   Ideate

8.4.1.1   Design of the PRISM Software

In the *exploration* phase of the design of the PRISM system, we generated ideas about the system and developed a prototype of it. Our initial ideas about what features should be available on the PRISM system were based on feedback from Carousel Information Management Solutions regarding the use of the features available on the Big Screen Live platform, data from other CREATE activities (e.g., focus groups on technology use; assessment of technology use patterns), and findings from our survey study. We decided that system features that were feasible and of potential value to our intended users included: internet, email, a resource guide, photo sharing, games, calendar, and a classroom feature. We were somewhat constrained by the existing Big Screen Live platform with regard to the number of features that could be placed on the system and extent to which existing features such as the games feature could be modified.

We generated ideas about the content for each of these features. For example, in the resource feature we decided that both national and local resources, as well as local events, should be included as tabs in the feature. Likewise, we discussed which specific resources should be included within each tab. For example, in the national resource feature we chose resources of relevance to older adults such as AARP and Administration on Aging, and for community resources we included local resources such as Area Agencies on Aging and transportation services. In the classroom feature we chose topics that would likely be of interest to older adults such as cognitive health, tips for traveling, and nutrition. We determined that the help feature should present both general system-level help (e.g., "how do I return to the home screen") and feature-level help (e.g., "how do I send an email?). The help feature was designed to present information in a procedural, step-by-step, "how to" manner (see Figure 8.4).

In addition, we discussed the design characteristics of the system, such as design and content of the home page, arrangement of the sidebar menu and the tabs within features, placement of the home and help buttons, and system login procedures. We considered aesthetic factors such as colors of the buttons, tabs, labels, background, and design of the PRISM icon. Our goal was to ensure that PRISM was useful, usable, and appealing. Our design decisions were based on current guidelines regarding interface design for older adults (e.g., Fisk et al., 2009), usability guidelines, the human–computer interaction literature, the literature regarding age changes in abilities, and data from our prior research such as our focus group on technology preferences and barriers to technology adoption (Mitzner et al., 2010).

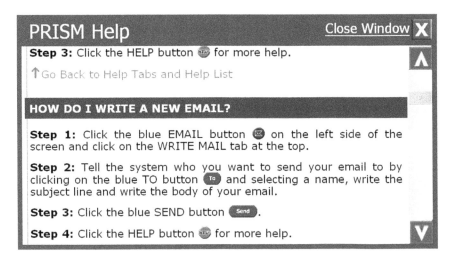

Figure 8.4 An example of a help screen.

8.4.1.2 *Design of the Instructional Materials*

We paid close attention to the design of the training program and instructional support materials. Our training protocol was based on the available training guidelines for older adults (e.g., Czaja & Sharit, 2013; Fisk et al., 2009). The training was spaced over three days so as not to overload the learner. Training included individual instruction and demonstration, active practice with immediate and specific feedback, and homework. Training began with basic computer, mouse, and windowing skills. Where possible we tried to make associations between new and familiar concepts. For example, to the extent possible we stressed the similarity between the computer keyboard and a typewriter keyboard. We then provided an overview of the PRISM system and the structure of the training program, followed by a gradual introduction to the features of PRISM. Each training session began with a review of the material covered in the prior session. We developed a user manual and easy-to-use brief "help" card. A usability expert initially vetted both of these documents. We designed the user manual to be compatible with the online help system. The content was structured in the same "how to" manner, and the color coding was consistent with that used in the software (see Figure 8.5).

8.4.2 *Prototype Development*

In this step of the process, we began to build a prototype of our initial conceptualization of the PRISM system. We then conducted two focus groups at the Miami site to gather initial feedback on the PRISM system. As discussed in Chapter 4, focus groups provided a way to gather qualitative

Figure 8.5 PRISM system help manual, table of contents.

information on perceptions, attitudes, and feelings in a group format, which can be more efficient than individual interviews. The focus groups included 14 adults (5 males and 9 females) aged 60–85 years ($M = 74.00$; $SD = 8.85$), all but one of whom had computer experience. The participants were introduced to the PRISM system and shown an early mockup via a PowerPoint presentation. We chose to use PowerPoint to introduce PRISM because the physical prototype was not yet fully operational. Presenting a system with glitches and limited functionality may have biased the perceptions of our participants.

Participants were asked to comment on the potential value of PRISM, the planned features and content of the features, and the interface. We gathered input on content for the classroom and resource features, the screen graphics, and the functionality of the features. For example, the participants indicated that the home page should include the time and date and that a tab for note taking should be included (e.g., grocery list) in the calendar feature. Ideas for content for the classroom feature included information on home safety and communication. Other comments focused on the screen graphics such as the readability and the choice of icons for each feature. Participants were asked to complete a system evaluation questionnaire. All of the participants (100%) indicated that they perceived PRISM as useful and that it would make their life easier and more productive. They indicated that the use of PRISM would better enable them to perform daily activities. They also indicated that they felt that

PRISM would be easy to use and learn. The PRISM prototype was refined based on the results of the focus groups.

8.5   Design of the PRISM System: Materializing

8.5.1   Usability Analysis

The materializing stage included a usability review of the working prototype, pilot testing of the prototype, and implementation of a cross-site randomized controlled field trial. Initially the CREATE team reviewed the prototype with respect to adherence to existing usability criteria (Czaja et al., 2019). Standard usability assessment tools such as heuristic analysis and cognitive walkthrough were used to identify potential user difficulties (see Chapter 4). A heuristic analysis allows usability problems to be identified early on in the design process so that problems can be resolved prior to a user's interaction with a system. The cognitive walkthrough evaluates the ease with which a new user can carry out a task using the system (e.g., send an email). Both of these tools are designed to enhance usability and ultimately user satisfaction and uptake of system or application.

Each screen of the PRISM system was reviewed to ensure the content was well organized, uncluttered, and easily visible (font and contrasts). The layout and formatting of information was reviewed to ensure it was consistent within and across features. This included, for example, placement of text in the header screens (e.g., consistent placement of the home button) and the color coding within each feature (e.g., the manner in which a tab was highlighted when activated). In addition, the functionality of each feature (e.g., email) and tab (e.g., sending a message) was reviewed to ensure consistency across features and tabs. The ease of performing tasks was examined. For example, it was noted that some of the games appeared to be cumbersome to perform. Given that the games were set, the instructions were modified to make them easier to follow. Several features of the calendar such as setting up a reminder were difficult to use and needed modification. The help feature required improvement to enhance usability.

8.5.2   Pilot Testing of the PRISM Prototype

Following the heuristic analysis and cognitive walkthrough, the PRISM system was again refined. Pilot testing was then conducted at all three sites. The sample included five adults per site (five males; ten females) who ranged in age from 66 years to 87 years ($M = 77$; $SD = 8.14$). The majority of the participants had some experience with computers (80%) and the internet (67%). The participants were trained on the system and observed by a trained research assistant while using the features of PRISM to perform sample tasks. They also completed an evaluation/usability questionnaire and a short interview regarding likes, dislikes, and needed improvements in PRISM. The majority (87%) of participants indicated that it was easy to learn how to use PRISM, that they were satisfied with PRISM as a whole (93%), and that it was enjoyable to use (80%). They also reported that each of the PRISM features was valuable. The participants provided important feedback on needed modifications to the system such as making the help system and the calendar easier to navigate and providing more training on basic window operations and use of the mouse.

Based on the results of the pilot testing, the PRISM system was further refined and pilot tested a second time at each site. We also tested our training protocols. In this round of pilot testing, we included four participants per site (four males and eight females; aged 67–87; $M = 75.08$; $SD = 6.22$). Feedback from this round of pilot testing resulted in further refinements to the training protocol and the system interface. For example, further refinements were made to the online help and calendar features. Participants indicated that it would be helpful to be able to practice using the computer mouse. A PRISM primer and mouse practice exercises were added to the classroom feature. We enhanced our training on the use of the printer. We realized it was important to reassure our participants at the beginning of training that if they made a mistake they would not "break" the computer and that if they experienced difficulties it was not due to their inability to learn but rather to the design of the system. We further emphasized that it takes time and practice to learn new skills.

All of the participants (100%) indicated that they found it easy to learn to use PRISM, that PRISM was easy to use, and that they were satisfied with PRISM overall. In addition, all of the participants (100%) indicated that they were satisfied with the amount and quality of training they received and that the practice exercises were helpful and sufficient.

We pilot tested the training protocol for the binder condition (described below in Section 8.5.3.1) with a different set of participants. Briefly, as noted below, participants in the binder condition received a notebook with content similar to PRISM. As PRISM was going to be evaluated in a large scale RCT, we wanted to ensure that participants who were randomized to the binder condition had the same amount of contact as that in the PRISM condition. Thus, people who received binders were also trained for over three days. Based on the pilot test, we found that the training materials for the binder condition were satisfactory.

Following the further refinement of the PRISM system and our PRISM training protocol, we conducted a second heuristic analysis to ensure that the system adhered to usability guidelines (e.g., Czaja et al., 2019) and that there were no glitches in the functionality. We evaluated the data collection features of the system. During the RCT, we collected real-time data on system usage patterns, and therefore we wanted to ensure that the system captured the data needed for the trial. We pilot tested the trial assessment battery to ensure that the measures were understandable, capturing the information we needed, and were not too burdensome.

The battery included a background questionnaire that assessed basic demographic information and self-ratings of health, measures of computer attitudes, technology/computer/internet experience, general technology acceptance, and computer proficiency. Given that we wished to evaluate if the computer skills of our participants improved as a result of using PRISM, we developed and evaluated a Computer Proficiency Questionnaire (CPQ; Boot et al., 2015). The questionnaire was designed to assess the proficiency of computer-based activities to support social engagement/support (communication), information gathering (internet), prospective memory (printing and scheduling software), and cognitive stimulation (multimedia use). In addition to these domains, the CPQ assessed the basic ability to interact with a computer using various input devices (computer basics).

The battery included a "life space" questionnaire that assessed mobility and activity patterns; a brief personality inventory; a measure of health literacy; a measure of general reading ability; measures of several cognitive abilities; a questionnaire that assessed various aspects of everyday memory functioning; and measures of social isolation, social support, loneliness, emotional well-being, and quality of life (Czaja et al., 2015). In addition to examining changes in our primary outcomes (social support, loneliness, social isolation) and secondary outcomes (computer attitudes, computer proficiency), our extensive battery allowed us to examine how these outcomes varied according to participant characteristics such as gender, age, and cognitive abilities. We tracked PRISM usage patterns and help requests. These data provide valuable information for future refinements of the system.

Participants in both conditions completed an evaluation questionnaire at both 6 and 12 months, which assessed satisfaction with PRISM or the binder. They completed a brief semi-structured interview regarding their overall impressions of how PRISM or the binder impacted their everyday activities.

8.5.3   Implementation: A Randomized Controlled Trial of the PRISM System

8.5.3.1   Overview of the Field Trial

As noted, we evaluated the PRISM system with a large and diverse sample of older adults at risk for social isolation who had minimal or no computer experience across three cities in the United States in a RCT. The sample included 300 participants (150 per study condition) who were enrolled in the trial. Participants were primarily female (87%), 64–98 years ($M = 76.15$, $SD = 7.4$), and ethnically diverse, and most were of lower socioeconomic status and did not have a college degree (78%). Participants had very limited or no prior computer experience and were at risk of being socially isolated, as defined earlier.

Participants were recruited from three cities in the southeastern United States (Miami, Atlanta, Tallahassee). Various methods were used for participant recruitment that included: advertisement in local media and newsletters, attendance at church and community meetings, interactions with agencies serving older adults (e.g., Meals on Wheels), posting flyers in senior housing buildings and public libraries, mailing lists, and participant registries.

Interested participants contacted the study site coordinator and completed a telephone screening for basic eligibility. Telephone pre-screening enhances efficiency and minimizes cost. Our baseline assessment was conducted in the homes of the participants; thus it was more efficient to conduct a telephone screening to determine if a participant met basic eligibility criteria (e.g., age, lived alone) prior to making a home visit. Prior to the assessment all participants completed a written informed consent for participation in the trial. A trained research assistant conducted the assessment, which included further measures of eligibility (e.g., vision and cognitive screening). We tracked how participants heard about the study and reasons for non-eligibility. This type of information helps to inform the design of future trials and is required for most academic journal publications.

Participants who remained eligible at baseline (e.g., met vision and cognitive criteria) were randomized into either the PRISM condition or the binder condition. Those assigned to the PRISM condition received a computer equipped with the PRISM software and a printer. The computer was installed in the participants' homes, and internet was provided for the study duration. Participants assigned to the binder condition received a notebook that contained content similar to that within the PRISM software. The binder contained a calendar, resource guide, games (e.g., word games, playing cards, and card game rule book), information about community groups, and information/tip sheets on the same topics as the "classroom feature" of the PRISM software. This material was updated monthly via mail. As described earlier, participants were trained over three days in their homes. We called them a week post-training to ensure that they were not encountering problems when using PRISM or the binder.

The duration of the trial was 12 months. Follow-up assessments occurred at 6 and 12 months post-randomization. We included "check-in calls" at 3 and 9 months, as well as a brief telephone assessment at 18 months. PRISM participants were allowed to keep the computer after study completion, and assistance was provided to help participants set up an internet account and the use of a browser as we could no longer support PRISM. Those in the binder condition had the option of receiving written materials on how to perform basic computer and windowing activities.

The trial was highly manualized. We developed a manual of operations (MOP) that included scripts for the screening, baseline, follow-up assessments, and the training sessions and standardized protocols for recruitment, assessment, implementation, and data transfer. A MOP is essential to ensure that the same procedures are followed at all three sites. The Institutional Review Boards at the site institutions approved the study protocol prior to any data collection activities. The trial included a Data and Safety Monitoring Committee, an independent group of experts who monitored the quality of the trial, recruitment, and participant safety. This committee was a requirement of the funding agency. The members of the committee provided valuable insights that helped improve the quality of the trial.

8.5.3.2 Main Findings

The findings from the trial were favorable. At six months participants who received PRISM reported significantly less loneliness and increased perceived social support and well-being. There was a trend indicating a decline in social isolation. Group differences were not maintained at 12 months, but those in the PRISM condition still showed improvements from the baseline assessment. There was an increase in computer self-efficacy, proficiency, and comfort with computers for PRISM participants at 6 and 12 months (Czaja et al., 2018). The increases in computer proficiency and computer attitudes were especially pronounced for participants who used the system more frequently. These participants also rated the usability of the system higher. Participants with higher cognitive abilities tended to use the system more frequently (Sharit et al., 2019). Moreover, individual differences in earlier use of the system, executive functioning, and computer efficacy predicted long-term use (Mitzner et al., 2019).

Most PRISM participants found it useful in their daily life (82%), indicated that PRISM made their life easier (80%), improved their daily life (84%), and enabled them to accomplish tasks more quickly (73%). They found PRISM easy (88%) and enjoyable to use (93%). The most valuable features were email, the internet, the classroom, and the games. PRISM participants reported that the use of PRISM made it easier to communicate with family and friends, engage in hobbies and play games, and look up community and health information. Email, the internet, and games were the most used features of PRISM. The use of the help feature declined over the course of the trial (Czaja et al., 2018).

8.6   Design Value

The PRISM experience clearly underscored the value of an iterative, user-centered design approach. We conceptualized, developed, tested, refined, and implemented the PRISM system. One measure of success was that we were able to train all of our PRISM participants on the use of PRISM. Our sample ranged in age from 65 years to 98 years, and 33% were 80 years or older. Further, the participants had very limited or no prior computer experience. The participants commented that PRISM was easy to use and learn; they rated the system as valuable and reported that it made their daily lives easier.

We learned a great deal from our pilot testing of the system and the training materials beyond what we learned in our rigorous heuristic analysis. The results of the pilot testing resulted in significant improvements to the system. Of particular note was the importance of pilot testing the training program. We learned that our participants needed more training on the use of the mouse and windowing operations. Had we not altered our training to account for this, participants may have struggled when attempting to use PRISM, which may have in turn negatively biased the results of the trial. Our pilot testing reinforced the idea that people need reassurance when learning to use new technology so that they are comfortable practicing the features and functionality.

The collection of the usage data provided valuable information. Along with the post-study interviews, these data provided insights into which features were most valuable. Overall, a multi-pronged assessment approach yields different types of information about a system. Some of the features such as the calendar and classroom feature were used less frequently. This usage pattern is likely due to the fact that despite improvements to the calendar during the design process, it was still a bit awkward to use. The classroom feature was only updated monthly, so infrequent use of this feature was not surprising. These findings underscore the fact that for a system or application to be used it must have perceived value and be usable. Our findings suggest that some aspects of this type of system should be dynamic so that users can discover new challenges and learn new skills. Many of our participants wanted to go beyond the features available in PRISM.

Given our target population, recruitment was a challenge as our focus was on individuals who were socially isolated. We had to use a multi-pronged recruitment approach. When conducting these trials, it was important to develop a systematic recruitment strategy prior to trial implementation and to allocate sufficient resources to recruitment efforts. The development of the MOP was also an important part of the process as it facilitated the implementation of the trial across sites and served as an important training tool for our research assistants. Finally, sufficient time must be allocated to system development and to the development of the implementation protocol. Evaluating systems in the field presents different challenges than evaluating systems in controlled environments.

8.7   Conclusions

Our intent in this chapter was to provide an integrated case study of the design and implementation of a technology system design to foster social engagement among older adults. We used a user-centered design approach and adopted the design thinking process. We employed many of the methods and tools described throughout this book. Overall, we learned a great deal from the PRISM trial. Users provide invaluable insight regarding the usefulness and usability of a system. These insights are important to incorporate into the design of a system. As we have stressed throughout this book, usefulness and usability are major factors driving technology adoption.

References

Berkowsky, R. W., Cotten, S., Yost, E., & Winstead, V. (2013). Attitudes towards and limitations to ICT use in assisted and independent living communities: Findings from a specially-designed technological intervention. *Educational Gerontology*, *39*(11), 797–811. doi:10.1080/03601277.2012.734162.

Boot W. R., Charness, N., Czaja, S. J., Sharit, J., Rogers, W. A., Fisk, A. D., Mitzner, T., Lee, C. C., & Nair, S. (2015). Computer proficiency questionnaire: Assessing low and high computer proficient seniors. The Gerontologist, *55*(3), 404–411, doi:10.1093/geront/gnt117.

Choi, M., Kong, S., & Jung, D. (2012). Computer and internet interventions for loneliness and depression in older adults: A meta-analysis. *Healthcare Information Research*, *18*(3), 191–198. doi:10.4258/hir.2012.18.3.191.

Cotten, S. R., Anderson, W. A., & McCullough, B. M. (2013). Impact of internet use on loneliness and contact with others among older adults: Cross-sectional analysis. *Journal of Medical internet Research*, *15*, e39. doi:10.2196/jmir.2306.

Cotten, S. R., Ford, G., Ford, S., & Hale, T. M. (2014). Internet use and depression among retired older adults in the United States: A longitudinal analysis. *The Journals of Gerontology Series B: Psychological Sciences and Social Sciences*, *69*(5), 763–771. doi:10.1093/geronb/gbu018.

Cudjoe, T. K. M., Roth, D. L., Szanton, S. L., Wolff, J. L., Boyd, C. M., & Thorpe, R. J. (2018). The epidemiology of social isolation: National health and aging trends study. *The Journals of Gerontology Series B: Psychological Sciences and Social Sciences*, *75*(1), 107–113. doi:10.1093/geronb/gby037.

Czaja, S. J., Boot, W. R., Charness, N., & Rogers, W. A. (2019). *Designing for older adults: Principles and creative human factors approaches* (3rd ed.). CRC Press.

Czaja, S. J., Boot, W. R., Charness, N., Rogers, W. A., & Sharit, J. (2018). Improving social support for older adults through technology: Findings from the PRISM randomized controlled trial. *The Gerontologist*, *58*(3), 467–477. PMCID: PMC5946917. doi:10.1093/geront/gnw249.

Czaja, S. J., Boot, W. R., Charness, N., Rogers, W. A., Sharit, J., Fisk, A. D., Lee, C. C., & Nair, S. N. (2015) The Personalized Reminder Information and Social Management System (PRISM) Trials: Rationale, methods, and baseline characteristics. *Contemporary Clinical Trials*, 40, 35–46. doi:10.1016/j.cct.2014.11.004.

Czaja, S. J., & Sharit, J. (2013). *Aging and skill acquisition: Designing training programs for older adults*. CRC Press.

Dickinson, A., & Gregor, P. (2006). Computer use has no demonstrated impact on the well-being of older adults. *International Journal of Human Computer Studies*, 64(8), 744–753. doi:10.1016/j.ijhcs.2006.03.001.

Fisk, A. D., Rogers, W., Charness, N., Czaja, S. J., & Sharit, J. (2009). *Designing for older adults: Principles and creative human factors approach* (2nd ed.). CRC Press.

Hawkley L. C., Kozloski, M., & Wong, J. (2017). *A profile of social connectedness in older adults*. AARP.

Mitzner, T. L., Boron, J. B., Fausset, C. B., Adams, A. E., Charness, N., Czaja, S. J., Dijkstra, K., Fisk, A. D., Rogers, W. A., & Sharit, J. (2010). Older adults talk technology: Their usage and attitudes. *Computers in Human Behavior*, 26(6), 1710–1721. doi:10.1016/j.chb.2010.06.020.

Mitzner, T. L., Savla, J., Boot, W. R., Sharit, J., Charness, N., Czaja, S. J., & Rogers, W. A. (2019). Technology adoption by older adults: Findings from the PRISM trial. *Gerontologist*, 59(1), 34–44. doi:10.1093/geront/gny113.

Sharit, J., Moxley, J. H., Boot, W. R., Charness, N., Rogers, W. A., & Czaja, S. J. (2019). Effects of extended use of an age-friendly computer system on assessments of computer proficiency, attitudes and usability by older non-computer users. *ACM Transactions on Accessible Computing*, 12(2), 1–28. doi:10.1145/3325290.

Slegers, K., Van Boxtel, M. P., & Jolles, J. (2008). Effects of computer training and internet usage on the well-being and quality of life of older adults: A randomized, controlled study. *The Journals of Gerontology Series B: Psychological Sciences and Social Sciences*, 63(3), P176–P184. doi:10.1093/geronb/63.3.p176.

Additional Recommended Readings

Czaja, S. J., & Sharit, J. (2013). *Designing training and instructional programs for older adults*. CRC Press, Taylor & Francis Group.

Gitlin, L., & Czaja, S. J. (2015) *Behavioral intervention research: Designing, evaluating and implementing*. Springer Publishing Company.

chapter nine

Emerging Challenges and Approaches

The one constant with respect to technology is change. The rapid advancement of technology, including changes to existing systems and the emergence of radically new systems, highlights the continuous need for a user-centered design approach that involves older adults. Today technology evolves rapidly, and as it does, so do design considerations. This chapter is intended to be forward-looking, presenting a discussion of the rapid advancement of technology, emerging technologies and potential considerations for their design, how design principles for older adults may or may not change over the next decades, emerging methods to evaluate design, and the need for designers to anticipate the future.

9.1 Will Design for Older Adults Continue to Be Relevant?

In 2010 in the United States, 43% of older adults were online. Five years later, that number rose 20 percentage points to 63% and has continued: by 2019, 73% of older adults were online. If this general trend continues (internet usage increasing 10–20 percentage points approximately every five years), in the next 10 years the digital divide – at least with respect to internet usage – might not exist. This observed increase in internet adoption is the result of a combination of two factors: some older adult non-internet users are adopting the technology and younger adults are aging into the older adult category and bringing their technology experience with them. In less than five years, smartphone ownership among older adults rose from 30% in 2015 to 53% in 2019, and there is no reason to suspect that smartphone ownership will not continue to rise among older adults (especially as fewer manufacturers market non-smartphones). Further, over the past few decades older adult cohorts have reported less technology anxiety and greater technology self-efficacy (Lee et al., 2019). At various points in this book we have emphasized how differences in technology experience, proficiency, and attitudes between younger and older adults are important considerations for the design of systems and instructional support. But what happens when today's current technology-savvy younger adults advance into older age? Will there still exist an age-related digital divide to consider with respect to technology experience, attitudes, and proficiency when designing technology systems? And will future older adults still lag in their adoption of new technology in the future?

For a variety of reasons, it is our contention that these issues will still be important to consider when designing for future generations of older adults. First, with new technology comes new learning. As mentioned in Chapter 1, older adults can take substantially longer to learn novel systems compared to younger adults. New systems that represent incremental changes in existing technology may not be associated with greater learning costs for older adults compared to younger adults due to older adults' ability to leverage existing knowledge (Charness et al., 2001); however, for radically new technology systems that are vastly different compared to previously learned technologies, the cost of new learning could be substantial. Age-related differences in learning have the potential to maintain a digital divide for emerging technologies even as technologically proficient younger adults age into older adulthood.

Evolving and emerging technologies highlight the need for continued involvement of older adults in the design process as systems place new or different demands on the user. There is no reason to suspect that age-related cognitive and physical changes will be eliminated in the decades to come, meaning that future older adults will likely experience greater usability challenges with future systems and that considering age-related changes in the design of these systems will continue to be important. It is challenging to anticipate what a computer will look like in 60 years. Will future computers be as different from the pocket-sized, touch-screen computers we interact with today as today's smartphones are from the room-sized supercomputers of the 1960s? Whatever these systems eventually look like (perhaps the best source to turn to in anticipating this might be science fiction), they will nevertheless place some demands on the user. With age comes changes to perceptual, cognitive, and motor abilities that can influence an older adult's ability to meet these demands. This was true in the past, is true today, and will very likely to be true for future systems. Design for older adults will continue to be relevant.

9.2   Technology Trends and Emerging Technologies

Good design for older adults can be a challenge because the design process must be aimed at a moving target. Existing technologies change form, and new technologies emerge over time. It is useful to spend some time thinking about not just how to design for current technologies but also what design challenges might be anticipated in the future, and the potential negative consequences of rapid technological change for older adults today and in the future.

9.2.1 Technology Advances

The forward march of technology can be quantified in many ways. Moore's Law provides one way to measure the advance of technology by describing the exponential increase in computing power over time relative to the increase in the number of components that can fit on a microprocessor. In the 1960s, while serving as director of research and development at Fairchild Semiconductor, Gordon Moore was asked to predict the future of computing. He observed at the time that the number of transistors that could fit on a microchip doubled every year and predicted that this trend would continue over the next decade. Ten years later, he revised that estimate: the number of transistors that can fit on a microchip would double every two years, resulting in a doubling of processing power every 18 months. This pattern has generally held true to date, although forecasters predict that the exponential growth in computing power may be reaching an end soon as we approach physical constraints on miniaturization. Regardless, smaller, more powerful microprocessors have been responsible for many of the rapid advances in technology we have seen over the past few decades, resulting in handheld devices that now have as much computing power as the most powerful computers decades ago. The advance of technology can also be quantified by the rate of technology diffusion in society, and there is evidence that technology diffusion has also accelerated. While it took over 70 years for the telephone to reach 50% of U.S. households, it took only 14 years for the cellphone and 6 years for the MP3 player. These trends paint a picture of a rapidly and radically changing technological landscape. If these technologies are not designed to be inclusive, many individuals will be at risk for rapid exclusion from their benefits as well as from an increasingly technology-driven society.

As technology evolves, some systems change incrementally. For example, the evolution of the desktop computer to a laptop form factor represented an incremental change. Although the smaller form presented some design challenges to be solved, particularly for older adults (e.g., smaller screen size), these issues could fairly easily be anticipated, and existing design guidelines for older adults could be applied or adapted. The nature of the task performed (e.g., word processing) did not fundamentally change whether that task was done using a desktop or laptop personal computer. However, other technological advances are more radical. Take for example autonomous vehicles, which may be on the road in the next decade. Autonomous vehicles (particularly vehicles that incorporate higher levels of autonomous control) fundamentally change the nature of the driving task from an active control and navigation task to a passive monitoring and vigilance task (e.g., monitoring for automation failures). The technology places new demands on the driver compared to non-autonomous vehicles, and currently researchers are trying to understand how to design systems to allow drivers of all ages to best meet these demands and develop guidelines for the design and implementation of such technologies. Radically new technologies deserve special consideration in the study of how best to design for older adults as novel and perhaps unanticipated demands may be placed on the user.

9.2.2 *Anticipating Future Design Challenges*

A few trends are worth consideration in terms of anticipating upcoming design challenges. The model of healthcare in the United States and other countries continues to shift responsibilities for care to older adults themselves and their informal caregivers. This shifting of responsibilities will become increasingly necessary as the number of older adults in the world increases, and there is a shortfall of formal caregivers to address their healthcare needs. Technology to support the practice of healthcare at home is a growing area of interest, and poor design has implications not only for system adoption but also the health and safety of older users. Design challenges in this domain include the inherent complexity of healthcare (especially for older adults who are likely to have one or more chronic conditions), how to design systems for individuals with varying degrees of health literacy, and how to present relevant health information differently as a function of the role of the person viewing it (e.g., older adult, informal caregiver, health professional). Emerging technologies include health coaching apps, electronic medical records, electronic pill dispensers, and various telehealth devices. Radically new technologies in this domain may soon include artificial organs and novel bionic implants. An important design challenge to consider in the design of technology-based healthcare solutions relates to infrastructure: how will these systems be designed, for example, considering that many rural communities lack high-speed internet required for some telehealth solutions? Another important consideration in the domain of healthcare is privacy, including who can view health information and what information they can view, which can have important implications for technology attitudes and adoption.

In the home, emerging technologies include digital home assistants, smart home devices, and simple home robots (e.g., robotic vacuum cleaners). In the near future, the Internet of Things (IoT), home sensor systems, and more advanced personal assistance robots are likely to play a larger role in the home. These technologies, if designed well, have the potential to help older adults remain independent in their own homes longer. However, to be successful, the design process needs to consider not only the variability of users in terms of their attitudes, abilities, and preferences but also variability in their living contexts. Differences in living contexts (e.g., size and type of home, room layout, household clutter, number of individuals living there) need to be anticipated and designed to promote successful adoption and use of home technologies. Here too, the issue of privacy is a crucial one for technologies that measure and monitor activities within the home. Cybersecurity threats come into play as well, especially for devices that rely on software maintained by third parties outside of the designer's control, which may expose users' data that can then be exploited. Designers of smart home devices and home digital voice assistants are currently trying to navigate these challenges with varying degrees of success.

The nature of work is changing as well, becoming increasingly reliant on one's ability to effectively and efficiently use technology systems (Czaja, Sharit, & James, 2019). And as a result of population aging and other personal and economic factors, in the past few decades the number of older adults in the workforce has been increasing and is projected to continue to increase in the future. According to the U.S. Bureau of Labor Statistics, in 1988 the domestic labor force participation rate of adults 65 years of age or older was 11.5%. That number is projected to increase to 21.8% by the year 2026. During that same period, labor force participation of adults 75 years of age or older will increase from 4.2% to 10.8%. Important emerging technologies in the workplace include cloud-based storage, collaboration, and virtual meeting tools, and work environments in the future may include to a greater extent radically new technologies such as artificial intelligence (AI) decision aids, augmented reality, brain–computer interfaces, and powered exoskeletons. Many of these new technologies may invoke new design considerations. For example, with AI an important issue is explainability – that is, when an AI agent recommends or decides a course of action, can the system provide adequate explanation to the user for why that decision was made? What information should be provided about the reasons for AI decisions, and how should that information be presented? The ability of AI to explain decisions has crucially important implications for the collaboration between humans and intelligent agents. Involving older adults in the design of these emerging technologies will be important to support the increasing number of older adults in the labor force.

The engagement in leisure, communication, and social activities has been shaped and will continue to be shaped by emerging technologies, and designers of these technologies need to consider older users to maximize their impact. Age-related changes can affect mobility, and many older adults live alone. Technology can provide opportunities for individuals to experience and interact with distant locations and people and to engage in fun and rewarding experiences within their own home, potentially offsetting the effect of limitations that some individuals experience with age. Currently emerging technologies in these domains include exergames, SmartTVs, online learning, video chat software, and virtual meeting spaces. These activities may soon be impacted by more radically new technologies including co-located virtual reality and augmented reality experiences that bring distantly located people into the same digitally constructed environment, telepresence robots that can allow individuals to move through and interact with distant locations, and social robots that can provide companionship. Many design considerations are unique to some of these technologies. For example, for a companion or social robot – what should the robot look like and sound like? What social activities should it support? And how should these technologies be designed, so they enhance and supplement human interactions rather than replacing them? The methods and tools outlined in this book can help answer these questions when considering older users of these technologies.

Finally, as alluded to previously, the domain of transportation and mobility is currently undergoing a major transformation as a result of emerging technologies. In some ways, older adults have the most to gain from this transformation, assuming that barriers to technology-based solutions can be minimized. As the population ages, the number of older drivers on the road will increase substantially in many countries. With advancing age, perceptual and cognitive abilities change, which makes driving less comfortable. Age-related increases in susceptibility to crash forces also can make driving more dangerous for older drivers when a crash occurs. This is true today and will likely continue to be true in the future. Advanced driver-assistance systems (ADAS) and ride-sharing apps are emerging technology-based solutions to promote safe mobility for life, and radically new technologies such as autonomous vehicles and smart and connected vehicles are also being designed and tested and may soon benefit older drivers. However, even for solutions such as autonomous vehicles that take over most aspects of the driving task, it is not certain whether differential age-related risk for crash injury or death will be eliminated unless the design of these technologies considers the needs, preferences, attitudes, and abilities of older adults. A research topic of great interest today is autonomous vehicle handover situations. At some point, an autonomous vehicle will encounter a situation that requires human intervention. How should the system be designed to gracefully transition control from the vehicle to the driver? If designers of handover interfaces do not consider older users and the design process does not involve older adults, older adults may still experience a greater crash risk relative to younger adults.

9.2.3 Defining a Research Agenda

The constant change of technology might dishearten some researchers studying how best to design for older adults. Intensive efforts can be devoted to understanding all the intricacies of the design of a piece of technology in order for that technology to be useful to and usable by older adults, only for that technology to soon become obsolete. The Center for Research and Education on Aging and Technology Enhancement (CREATE) has, over the past two decades, had to carefully navigate this issue and has adopted approaches to address it. First, in our research agenda, we have made sure that research questions, whenever possible, go beyond device-specific issues. That is, much of our research has been aimed at understanding *general principles* of design. This strategy helps ensure that findings have relevance to newer technologies even as older technologies become obsolete. Second, over the past 20 years, CREATE has tried to anticipate which emerging technologies might play a larger role in the future. For example, CREATE researchers have already examined issues related to the design of social robots and the acceptance of virtual reality solutions and autonomous vehicles well in advance of these technologies becoming mainstream. A focus on general design principles along with an awareness of and anticipation of how technologies might change in the near future can help define a research agenda that will continue to be relevant and beneficial to the growing older adult population as the technological landscape shifts.

9.3 Emerging Approaches

A number of methods and tools were presented within this book to understand older users' needs and to evaluate design: needs assessment, heuristic evaluation, focus groups, surveys, simulation, and performance modeling. Advances in technology will likely change the nature of these approaches and emerging approaches may become more common. Advances in the ability to simulate complex systems and environments using virtual reality, for example, will likely make simulation a more common and convenient method over time (Chapter 5). It is also possible that in the near future neuroimaging may become a more prominent method with respect to design evaluation. The field of neuroergonomics is not a new one (Parasuraman, 1998), but technological advances now allow some of these techniques to be more easily deployed and make them more appropriate for design-focused research involving complex systems (e.g., advances in ambulatory neuroimaging techniques such as mobile electroencephalography). Some of the aims of neuroergonomics are to provide measures of workload, vigilance, and fatigue to better understand the demands a system places on the user. Other aims relate to uncovering fundamental insights into the relationship between performance and brain activity that may be useful in developing design principles. Cheaper, mobile, and less intrusive eye-tracking technologies may also play a larger role in understanding the information processing requirements of systems. Other emerging approaches include the development of new models with which to predict human performance or the adaptation of existing models to be more relevant to new environments and situations. For example, Chapter 6 presented how movement times of younger and older adults in the interaction with a system can be accurately predicted by Fitts' Law. Newer models of movement control are currently being developed, adapted from Fitts' Law, to predict movement times in immersive 3D virtual environments (e.g., Deng et al., 2019). Some of these performance models do not yet consider how age influences model parameters, but this will be a necessary consideration for future research. As technology advances, so will the methods and tools available to study the design of new technologies.

9.4   Conclusions

This chapter attempts to preview the future: simultaneous with the aging of the world's population there will be a rapid advancement of technology and rapid diffusion of technology into society. These technological advances will affect all domains of living. There is no reason to suspect that in the future age-related changes in perceptual, cognitive, and motor abilities will be eliminated, meaning that as new technologies emerge, their design will still need to consider the older user and there will be continued value in including older adults in the design process. To help meet the challenges of designing for older adults in the future, technological advances may also produce new and useful methods and tools to evaluate design. This chapter also emphasized the benefit of being future-oriented in design research. Successful design of technology-based solutions requires anticipating the future to some degree, and engagement in research that, when possible, asks and answers questions about principles of design rather than how to design a specific device that may become obsolete. Such a research agenda, along with utilizing existing approaches described in this book as well as emerging approaches described in this chapter will help ensure that current and future the benefits of technology can be realized by older adults to support and enhance their independence, productivity, health, safety, social connectedness, and quality of life.

References

Charness, N., Kelley, C. L., Bosman, E. A., & Mottram, M. (2001). Word processing training and retraining: Effects of adult age, experience, and interface. *Psychology and Aging, 16*(1), 110–127. doi:10.1037/0882-7974.16.1.110.

Czaja, S. J., Sharit, J., & James, J. B. (Eds.). (2019). *Current and emerging trends in aging and work*. Springer.

Deng, C. L., Geng, P., Hu, Y. F., & Kuai, S. G. (2019). Beyond Fitts's law: A three-phase model predicts movement time to position an object in an immersive 3D virtual environment. *Human Factors, 61*(6), 879–894. doi:10.1177/0018720819831517.

Lee, C. C., Czaja, S. J., Moxley, J. H., Sharit, J., Boot, W. R., Charness, N., & Rogers, W. A. (2019). Attitudes toward computers across adulthood from 1994 to 2013. *The Gerontologist, 59*(1), 22–33.

Parasuraman, R. (1998). *Neuroergonomics: The study of brain and behavior at work.* Cognitive Science Laboratory.

Additional Recommended Readings

Rogers, W. A., & Mitzner, T. L. (2017). Envisioning the future for older adults: Autonomy, health, well-being, and social connectedness with technology support. *Futures, 87,* 133–139. doi:10.1016/j.futures.2016.07.002.

Index

Page numbers in **bold** denote tables, those in *italic* denote figures.